Lionel Smith Beale

Disease germs

Their supposed nature

Lionel Smith Beale

Disease germs

Their supposed nature

ISBN/EAN: 9783337112790

Printed in Europe, USA, Canada, Australia, Japan

Cover: Foto ©berggeist007 / pixelio.de

More available books at **www.hansebooks.com**

DISEASE GERMS;

THEIR

SUPPOSED NATURE:

AN ORIGINAL INVESTIGATION,

WITH CRITICAL REMARKS,

BY

LIONEL S. BEALE, M.B., F.R.S.,

Fellow of the Royal College of Physicians; Physician to King's College Hospital; and lately Professor of Physiology and of General and Morbid Anatomy in King's College, London.

COLOURED ILLUSTRATIONS.

LONDON:
J. CHURCHILL & SONS, NEW BURLINGTON STREET.
1870.

[*All Rights Reserved.*]

PREFACE.

THE views advanced in this volume have been entertained by me for some years. The facts upon which they are based have been referred to in several papers, and especially in my Report on the Cattle Plague to the Royal Commissioners, published in 1866. They have also been frequently mentioned in my lectures during the last six or seven years.

The general conclusions at which I have arrived were sketched briefly in the last course of lectures I delivered at Oxford by direction of the Radcliffe Trustees.

The mode of investigation adopted for the preparation of the specimens for examination under the highest magnifying powers is the same that I have employed for many years past, and has been described in detail in other works.

L. S. B.

GROSVENOR STREET,
June 20th, 1870.

TABLE OF CONTENTS.

	PAGE
Of Poisonous Matter	1
Of a Germ	10
Of Bioplasm	11

OF VEGETABLE GERMS.

Of the Yeast Fungus	14
Of a Single Yeast Particle, and of its Bioplasm or Germinal Matter, and Formed Material	15
Of the Production of the minute Yeast Germs	16

OF GERMS IN THE AIR.

Of the Detection of Vegetable Germs in the Air	22
The Characters of the most minute Vegetable Germs	33
Of Germs of different kinds of Vegetable Organisms	35
Of the Origin of Vegetable Germs. Spontaneous generation	37

SUPPOSED INFLUENCE OF VEGETABLE GERMS IN CAUSING DISEASE.

The Manner in which they might enter the Body	62
Of the Vegetable Germs actually discovered in the Fluids and Tissues of the Higher Animals during Life	63
Question of derivation of Fungus Germs from higher Bioplasm of another kind	71
Of diseases known to be due to Vegetable Organisms	74
Some difficulties in accepting the Vegetable Germ Theory of Disease	76

EXPLANATION OF THE PLATES.

Plate I.—Fig. 1. Development of germs in organic fluids, showing the changes from day to day.

Figs. 2 to 4. Bacteria germs under magnifying powers varying from 1,800 to 5,000 diameters.

Fig 5. Most minute germs, visible under the one-fiftieth of an inch object glass.

Fig. 6. Yeast cells growing, under powers of 215 and 400 diameters.

Fig. 7. Yeast cells with germs growing from them, magnified 1,300 diameters.

Fig. 8. Young growing yeast cells, under a power of 2,800 diameters.

Plate II.—Fig. 9. Old spores of fungi, with very thick formed material.

Fig. 10. Germs of fungi, showing relation of bioplasm to formed material.

Fig. 11. Passage of bioplasm through pores in the formed material, to illustrate formation of branching stems.

Fig. 12. Extremities of rapidly growing fungus stem, to illustrate rapid growth of bioplasm.

Fig. 13. Most minute bacteria × 2,800.

Fig. 14. Germs of bacteria, growing and multiplying rapidly.

Plate III.—Fig. 15. Vegetable organisms and other minute masses of bioplasm from surface of villus—Cattle Plague.

Fig. 16. Vegetable organisms from a clot of blood.

Figs. 17 to 20. Living germs from contents of hermetically sealed vessels.

Fig. 21. Vegetable growth in mucus.

Fig. 22. Vegetable growths in different stages of development in sputum.

Plate IV.—Fig. 23. Bacteria from intestinal contents—Cholera.

Fig. 24. Free bacteria germs from columnar epithelium—Cholera × 1,800.

Fig. 25. Bacteria in the contents of an obstructed capillary.

Fig. 26. Vegetable germs in blood of a cow which died of Cattle Plague.

Fig. 27. Sporules of fungi in contents of capillary.

GERMINAL MATTER AND THE CONTACT THEORY.

For a general statement of the nature of morbid poisons and the growth and changes occurring in germinal matter in disease, the reader is referred to a little work by Dr. Morris, entitled "Germinal Matter, and the Contact Theory: an Essay on the Morbid Poisons, their nature, sources, effects, migrations, and the means of limiting their noxious agency." CHURCHILL AND SONS.

CONCERNING MAGNIFYING POWERS.

In order that some idea may be formed of the degree of magnifying power of the *one-fiftieth of an inch object glass*, lately made for me by Messrs. Powell and Lealand, it may be stated, that if it were possible to see a human hair in its entire width under this power, it would appear to be nearly one foot in diameter, and an object an inch in height would be made to appear 250 feet high. A child of three feet would look about as high as Mont Blanc.

AN APPARATUS FOR COLLECTING PARTICLES SUSPENDED IN AIR.

Dr. Maddox has described a convenient arrangement for the purpose of collecting solid particles suspended in the atmosphere, in the Monthly Microscopical Journal for June 1st, 1870. This ought to have been referred to on p. 22; but, unfortunately, that part of my work was revised before I received the Journal.

DISEASE GERMS:

THEIR

SUPPOSED NATURE.

POISONOUS matter exists in different states. There are poisonous solids, poisonous liquids, and poisonous gases. These kill man and the higher animals in many different ways. Some destroy the textures of the body; others interfere with the due discharge of function on the part of certain tissues and organs, the proper action of which may be necessary to the living state; while others again actually kill the living matter of the body; and of these there are poisons which act as soon as they pass into the organism, almost instantaneously,—and poisons which produce their deleterious effect very slowly. There are again substances which act as poisons to *every known kind of living matter*, while certain principles of vegetable origin which destroy some vertebrate animals with unerring certainty may be eaten freely by others, and even constitute their ordinary food.

Of all the poisons which cause the death of man, the most subtle and the most difficult to isolate and investigate are those which give rise to certain forms of disease which spread from person to person. Many poisons of this class, as compared with non-living poisonous agents, act very slowly, and when they destroy life, do so, not by their immediate action upon tissue or living matter, but by their indirect influence upon the physiological changes going on in various tissues and organs. Many of these poisons are indeed uncertain in their action, and are for the most part fatal to a very small percentage of those attacked. They not unfrequently cause serious structural damage in consequence of which the organism becomes predisposed to certain forms of disease; and oftentimes, it may be not until some time has elapsed, although the individual invaded escapes with his life, some delicate organ is irreparably injured by the changes produced by the action of the poisonous material, and death follows after a varying interval of time.

These poisons not only seriously derange the healthy functions, but having entered the body they multiply many million fold. They are *living*, and increase as living particles alone increase. They grow, they feed upon the nutrient juices of the organism and upon the tissues, and in some cases flourish at their expense and destroy them. The poison when it enters may be so infinitesimal in quantity, that it can neither be measured nor weighed, nor under ordinary circumstances seen ; but having gained access to the blood and tissues, it increases to such an extent that in many cases sufficient is produced in one subject to infect hundreds of persons, the population of a town, or even a whole country.

The infecting matter has been supposed by some to consist of some subtle entity, which was not cognizable by the senses, or to be made evident in any way. By others it has been considered to be matter in a gaseous state. Great authorities have defined it, with apparent precision, to be a chemical body, which may exist as a volatile vapour,* in a solid state, or dissolved in fluid. With a still more decided appearance of precision, the contagious poison has been pronounced to be an *albuminoid matter* in a state of rapid chemical change, which has not yet been isolated. Thus, the Cattle Plague Commissioners expressed the opinion that the poison of that highly contagious malady was probably matter " of a kind which is, and always will be, undiscoverable by the microscope." They remark "that

---

* The view that some contagious poisons are *volatile* seems to have been adopted because it has been found that the disease could be propagated through air and by the breath of the affected man or animal, but such a conclusion is unjustifiable, for, as is well known, particles as large as starch globules can be wafted from place to place by currents of air, and these are not *volatile*. Moreover, multitudes of insoluble particles, of determinate size, always exist, buoyed up by the ordinary air on the surface of the earth. But solid particles should not be called *volatile* unless they can be converted into vapour which, by condensation, again assumes the general form and state the particles first exhibited.

chemistry has as yet (1866) found in Cattle Plague no complex albuminoid matter in a state of rapid chemical change, capable of communicating its own action to the albumen of the serum of the blood, and of the textures of cattle." The author of these sentences, unwittingly perhaps, expounds the doctrine he himself believes, and desires the public should adopt. Though he admits that chemistry has not found the matter, he intimates that ere long success may be attained. On the other hand, he has no hope whatever that the microscope will ever assist in the discovery of the poison. The poison is *always to remain undiscoverable* by this instrument! Here is one mode of investigation authoritatively condemned without reason and without knowledge on the part of him who condemns it, and another exalted with unmistakable affection and partiality, although its shortcomings are admitted. At the very time these sentences were written, the poisonous matter *had been made out* by the microscope, and the contagious material of more than one contagious malady had been actually figured in the report which is here criticised, and which the writer is supposed—for he does not hesitate to comment upon it—to have read and studied.

Those who are acquainted with the theories now entertained upon this very important and interesting question, know that many of them are based upon the results of microscopical enquiry. There are indeed no other means of investigation yet discovered which afford equal prospects of success. Moreover, by this method of enquiry important results have been already obtained. It was therefore unfortunate that microscopical investigation should have been condemned by the authority of a Royal Commission, and the suggestion offered that chemistry was likely to succeed in detecting particles of matter which it is only possible to discern by the aid of very high magnifying powers employed with due care.

Notwithstanding the publicity given to many mere physico-chemical theories and the energetic means taken to force them into popular favour, the view that the poisonous matter of self-propagating (Dr. Budd) diseases, really consists of some kind of living substance (germinal or living matter), has been steadily gaining ground, although it must be admitted that many of the statements which have been made in its favour are of the vaguest and most unsatisfactory description. Thus many persons seem to have adopted the strange notion that dust is made up of disease germs. The harmless ingredients of this material have been spoken of as if they were disease-propagating particles of the most terrible kind. The innocent organic fragments detached from our carpets and rugs and dresses and furniture, have been treated as if they were contagious.

Of those who accept the *germ theory of disease*, some hold that the living particles which grow and multiply in various kinds of decomposing organic matter, are derived not from germs floating in the air, but have been built up *ab initio*—result in fact from spontaneous generation or heterogenesis. These think to fortify their position, which indeed sorely needs even an appearance of strength, by such remarks as the following:—" There is not a physiologist of eminence who would deny the possibility of the origin of organic forms direct from the inorganic, while many are convinced of the truth of the doctrine," and so on,— as if this sort of statement could in any way influence the judgment, or help the elucidation of the truth as regards the question at issue.

On the other hand, it is not uncommon to find writers in our journals affirming with the utmost confidence, that these so-called *germs* are fictions of the imagination, that they have never been seen, and that the air, which by some is said to teem with disease germs, is entirely free from them.

Some authorities have opposed the germ theory of disease by the argument, that if the supposed germs really did exist, they would certainly exhibit well-marked distinctive characters. But there is no ground whatever for the opinion that the particles which are instrumental in propagating contagious diseases, if living, should exhibit specific structural peculiarities. That the several kinds of contagious matter of small-pox, scarlet fever, and other eminently contagious maladies, are distinct, each being imbued with its own specific and peculiar power, is demonstrated by observation, but that this difference in power or property should be associated with any recognizable difference in appearance, form, chemical composition, or any physical characters, is exactly what a careful examination of the facts, and a judicial weighing of the evidence, would not lead us to anticipate. If, indeed, these poisons really consist of living matter, analogy would lead us to conclude that they would not be distinguishable from one another, except by the effects they produce. For we cannot distinguish one form of healthy living matter from another. The living matter which produces nerve, or muscle, or bone, is just like—at least as far as we can ascertain—that which gives rise to cuticle or to cartilage? Though the results of the life of these several kinds of living matter are so very different, the matter itself appears similar in all cases; and, examine it as we may, we cannot discover any distinguishing marks. Nay, the living matter of one animal is like that of another. No form of animal living matter can be distinguished from the living matter of a plant, or from that of any of the lowest simplest forms of existence. Discarding the results of observation, some writers have indeed maintained that the anatomical elements or cells of morbid growths have peculiar characters of their own, and can invariably be distinguished from those of healthy structures; but repeated failure of the attempts to do so on the part of those who held that the distinction existed, necessitates a modification of this view. But even if characteristic differences could invariably be pointed out in the anatomical units or cells, it is quite certain that the *living matter* of the most virulent cancer, or of the most inveterate contagious malady could not be distinguished from that of the harmless growth, or healthy tissue at any period of its existence.

The old doctrine, that the disease-carrying particles, are germs derived from and capable of producing microscopic vegetable organisms, has been recently revived and extended. Since the investigations of Pasteur, and the publication of the observations of Salisbury and others, the *vegetable* or *fungus germ theory of disease* has received a large accession of advocates. According to this view, there are different species or varieties of fungi corresponding to the several contagious maladies with which we are familiar.

We must not, however, conclude that if disease germs really do consist of living bioplasm or germinal matter, they must necessarily be of a vegetable nature and have sprung from vegetable organisms, or have originated spontaneously, for it is obviously possible that, though *living*, their nature may be very different, and that they may have been derived from a different source. While I freely admit that the facts of the case are conclusive as regards the living state of the active matter of contagious diseases, I am quite unable to subscribe to the arguments advanced in favour of the *Vegetable Germ Theory of disease.*

But it is time we should pass on to consider what is the nature of the supposed *germs* to which contagious maladies are now attributed. If the opinion is very generally entertained that the *materies morbi*, the *virus* or *contagium* of contagious diseases consists of germs which are introduced into the organism, the exact nature of the germs in question, at any rate is the subject of much discussion, and, indeed, the nature of germs generally, as well as the question of origin of these bodies, and the manner in which they act has given rise to many different theories. And surely it is a point not only profoundly interesting, but of vast practical importance, and worth any effort, to determine, whether the germs upon which the communicability of contagious diseases alone depends, are certain species of low simple organisms of definite character, produced in the outside world independently of man, or are bodies which have originated in man himself, and are to be regarded as degraded forms of living matter, derived by direct descent from some form of human living germinal matter or *bioplasm*. If the first supposition—and this is the favourite doctrine at this present time—should turn out to be correct, there appears much less hope of extirpating the diseases the germs produce, than if the last-mentioned theory, or some modification of this, should turn out to be true.

Of a Germ.—The term *germ* can only be correctly applied to a particle that is *alive;* but there are multitudes of different kinds of germs. It used to be supposed that the germ grew in a manner peculiar to itself; but we now know that however varied may be the substances resulting from the changes taking place in germinal matter, every kind of this living material, at every stage of existence, *grows* essentially in the same manner, though at a very different rate. The living particle which sprouts from a cell of the adult plant or organism, and is then detached, may be called a *germ*, as well as the living particle formed in the ovum, or the living matter in the ovary from which the new being is evolved. Any living particle growing or capable of growth, may be termed a "*germ.*" Every *germ* comes from living or germinal matter, and from this only. A particle of living matter, less than the $\frac{1}{100,000}$th of an inch in diameter, is a living germ, Figs. 1 to 8, plate I. It may take up lifeless matter, and convert this into living matter like itself, and thus *grow*. It may then divide and sub-divide, and thus a mass of considerable size may result. The original *germ* may give rise to successive generations of new particles, "germs" having similar powers or properties; or from it may emanate higher types of organization, having special formative powers, from which new germs may or may not proceed. So that a *germ is but a particle of living matter, which has* been detached from already existing living matter, and this living matter came from matter of some sort which lived before it.

Of Bioplasm.—Hitherto, I have employed the simple term *germinal or living matter*, to denote the active matter which is alone instrumental in the formation of all living beings and their tissues and organs; but this term is lengthy, and, in some respects awkward, and inconvenient. It cannot be used alone when speaking of a single particle, nor can it be employed adjectively. The word *Protoplasm* has been much used for some years past, but the vagueness attached to it is fatal to its employment here. A word is wanted to denote living, forming, growing, self-producing germinal matter, as distinguished from matter in every other state or condition whatever. Now protoplasm has been applied, both in this country and in Germany, to *lifeless* matter as well as to *living* matter, to *formed matter and tissue* as well as to the *formative* matter. And quite recently, Prof. Huxley and others have added to the confusion by giving it a still wider signification—so very wide, indeed, that almost anything that ever formed part of an organism may be called protoplasm. Roast mutton, white of egg, and a number of other things living and dead, having structure, as well as structureless, are said to consist of protoplasm; so that the word may include almost anything, and is not applied to matter in any particular state. It becomes, in fact, useless. The term I propose to apply to the *living* or *germinal self-propagating* matter of living beings, and to restrict to this, is *Bioplasm* ($\beta\iota o\varsigma$, life ; $\pi\lambda\alpha\sigma\mu a$, plasma). Now that the word *Biology* has come into common use, it seems desirable to employ the same root in speaking of the matter which it is the main purpose of biology to investigate. *Bioplasm* involves no theory as regards the nature or the origin of the matter. It simply distinguishes it as *living*. A living white blood-corpuscle is a mass of bioplasm, or it might be termed a *bioplast*. A very minute living particle is a bioplast, and we may speak of living matter as bioplasmic substance. A cell of epithelium consists of *bioplasm* or bioplasmic matter, surrounded by *formed non-living* matter, which was however once in the *bioplasmic state*. In the same way an oval yeast particle consists of the *bioplasm* with an envelope of *formed material*, which has resulted from changes occurring when particles upon the surface of the bioplasm died. The bioplasm of the microscopic fungus or other organism may give off diverticula which may become free independent *bioplasts*. Each minute bioplast may grow, and in the same way give rise to a number of other bioplasts.

The mode of growth of the *bioplasm*, and the manner in which it undergoes conversion into *formed material*, will be understood if Figs. 9 to 12, plate II, p. 20, be attentively examined. In Fig. 12, the bioplasm is growing very rapidly at the extremity of each of the branches of microscopic fungus figured. Here it has imbibed much of the carmine employed to stain it, and thus render it distinct, and is coloured with the greatest intensity. The formed material in this situation is thinner than elsewhere, sufficient time not having yet elapsed for it to become of the ordinary thickness which it exhibits in that part of the branch which is fully formed. The formed material is thickened by deposition, layer within layer, in the manner shown in Fig. 9 *a*, and as will be again referred to further on. (See page 16.)

In order to form a general notion of the nature and properties of living germs which consist partly or entirely of *Bioplasm*, it will be necessary to consider carefully what takes place when a very simple organism grows. I propose, therefore, in the first place to discuss the question of the formation of the yeast fungus or yeast cell. The organism is so common that any one can easily obtain it and study for himself the phenomena which will be referred to. This will form a good introduction to the consideration of the *vegetable germ theory of disease*, which will clear the way and prepare us for the full consideration of the highly interesting but complex question of the nature and origin of the contagium, or virus, of contagious or self-propagating diseases.

Of the Yeast Fungus.—If on a warm summer's day or in a warm room a piece of germinating yeast about the size of a pin's head be placed upon a glass slide, covered with very thin glass and gently pressed, so as to form a very thin layer, and then examined under a microscope having a magnifying power of about 300 diameters or upwards, numerous little transparent, colourless, oval bodies will be seen all over the field. The bioplasm of these may be coloured by carmine (Figs. 6, 7, 8, plate I.). They vary in size much more than is generally represented in published drawings of them, and many are joined together. If the stratum of yeast be made very thin, as may easily be effected by firm even pressure upon the thin glass, numerous very minute bodies will be observed amongst the well marked yeast corpuscles. The yeast consists of the distinct oval bodies, the minute particles (detached germs), and a fluid in which both are suspended.

If now the apparently smooth oval bodies be subjected to examination under a still higher power (1,000 diameters, or upwards), many of them will be found to exhibit little eminences, which project from various parts of the surface. Sometimes only one such eminence, and sometimes as many as a dozen, may be counted in connection with one oval yeast cell (Fig. 7, plate I.). These are undetached germs;

and if the yeast cells be kept under the field of the microscope for some hours at the proper temperature, and then examined again, the little projections already adverted to will be found to have increased in size, and *some will have detached themselves from the parent yeast cells.* Thus the minute bodies (germs) suspended in the fluid portion of the yeast have arisen from yeast cells which existed before them. The little eminences or projections are, in fact, buds or germs formed by the yeast cell, and when detached, constitute new yeast plants, and thenceforth become independent organisms. Each germ or bud was for some time connected with the parent cell, from which it originated. The very material of which it consists was continuous with that of the parent. The matter of which it was formed, in fact, formed a part of the parent mass, as is well seen in Figs. 7 and 8, plate I., page 16. Once detached, however, each little germ becomes an independent living thing, and never joins again the particle from which it emanated. Many of these germs detached are represented in Figs. 7 and 8. The free germ may multiply while it remains in the *germ stage of existence*, or it may grow into the likeness of its parent if supplied with the proper nutrient material and placed under conditions favourable for its advance to a higher stage of its development.

Of a single yeast particle and of its Bioplasm and formed Material.—If a single yeast cell be carefully examined it will be found to be composed of two different kinds of matter, the one smooth, *transparent* and *external*, forming a membrane closed at all points, and commonly known as the *cell wall* (formed material), the other *soft, diffluent*, also *transparent*, but apparently composed of *semi-fluid* matter (*germinal matter or bioplasm*). If the yeast cell be firmly pressed between glass plates it will burst, and the contents will be squeezed out as soon as the rupture of the protecting envelope has been effected. This envelope or cell wall varies in thickness in different cells, being firm and thick in the oldest, and so thin as to be demonstrated with difficulty in the youngest cells. The envelope itself is formed not by the deposition of matter from the surrounding medium upon the surface of the cell, but from the soft diffluent bioplasmic matter within. This envelope is thickened by the formation of new formed material from the bioplasm which is deposited layer within layer upon the inside of the already formed capsule, cell wall, or envelope (Fig. 9 *a*, plate II.). The germinal or living matter or bioplasm, can be coloured with an ammoniacal solution of carmine, while the formed material remains perfectly colourless. This fact is shown in the figures which have been coloured so as to resemble the specimens from which they have been carefully copied (see Plate I., figs. 6, 7, 8; plate II., figs. 9 to 12).

Of the production of the minute Yeast Germs.—Now, how are the little *buds* or *offsets* or *gemmules* of the

GERMS.—FUNGI PLATE I

Fig. 1.

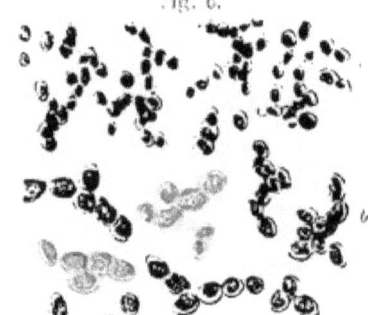

Fig. 6.

development of germs in organic fluid. a. ... first appearance. b. one day ... days. d. four days. e. five, and f. six days. × 300. p. 10.

... a × 315. b × 300. p. 14, 10.

Fig. 7.

Fig. 2.

...germs from the ... × 1500. p. 34.

Fig. 3.

The same. × 3000. p. 34.

Fig. 4.

The same. × 6000. p. 34.

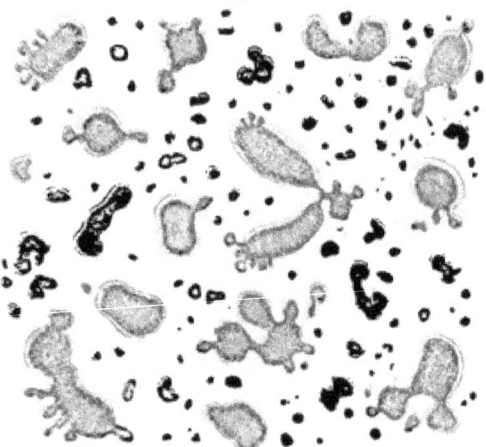

growing yeast cells, showing diverticula for each mass of the bioplasm. These are, from time to time, detached each germ when set free may grow and produce others × 1300. 1807. p. 14, 19.

Fig. 8.

Fig. 5.

The most minute germs of fungi visible under the 1/16. The smallest is much less than the 1/160000 of an inch in diameter. p. 34.

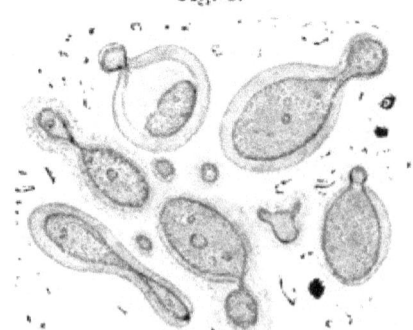

showing yeast cells and most minute germs, well stained with carmine and magnified by the 1/16 == 1300 diameters. p. 14. Nov., 1869

1/1000 of an inch ——— × 615 lin. ear.
1/10000 " " ——— × 1500.
" " " ——— × 3000.
" " " ——— × 6000.

yeast plant formed? Do they result from the growth of the external envelope with which they are certainly for a time connected, or is the soft diffluent material within concerned in their production? This question can be answered conclusively from direct observation. Never do we find a bud which does *not contain some of the soft diffluent germinal matter; never one composed of the matter of the envelope only* (Figs. 7, 8, Plate I.). In the formation of these buds a very small portion of the soft material protrudes through minute pores in the envelope, perhaps pushing a very thin layer of this latter before it, through which it imbibes nutrient matter from around. It soon increases in size forming a little nodule or button which remains for a time connected with the parent mass by a very narrow pedicle (Fig. 8, Plate I.). This pedicle consists externally of matter like that of which the envelope or cell-wall is composed, but in its centre may be traced a very thin line of bioplasm, by which the bioplasm or germinal matter of the bud remains for a time connected with the parental bioplasm. This line cannot be discerned in every instance, but from the numerous observations I have made I do not believe it is ever absent at this period of growth. I have seen the diverticulum of germinal matter in so many cases already projecting from the general mass, that I feel sure the above description is correct. When multiplication is going on with great rapidity, the mode of formation of the buds may in fact be demonstrated with the greatest certainty, as represented in Fig. 8, plate I.

The material of which the cell wall consists (formed material) passes gradually into the germinal matter, when the plant is germinating quickly, and the abrupt line, which marks the internal boundary of the capsule in many specimens, is absent. These points are given in the drawings, Figs. 7 and 8, the latter, in which the formed material is made too thick, taken from a specimen magnified by the fiftieth.

Where growth is active, the yeast cells are embedded in a soft material continuous with the external surface of the envelope, as represented around the yeast particles in the central part of Fig. 8. This probably consists partly of matter which is drawn towards the surface of the cells by the currents of fluid which are setting towards the germinal matter within, and partly of imperfectly hardened formed material. Thus there always appears to be a space between the outer part of one and the outer parts of neighbouring cells.

The very soft material, which consists of imperfectly formed matter, that by gradual condensation assumes all the characters of the cellulose wall of the yeast particle, corresponds to the mucus which lies between the particles of bioplasm concerned in the formation of that substance, and bears the same relation to the envelope of the yeast cell as the viscid mucus does to the wall of an epithelial cell embedded in it.

When yeast has been successfully stained with carmine fluid it forms a beautiful object for investigation, and with the aid of very high powers many points of the greatest interest may be discovered. The cells represented in Fig. 7, plate I., were growing very rapidly. Protrusions or outgrowths are seen projecting from every part of the bioplasm, and in some instances the latter appears to have been preserved while it was in the very act of moving. I have specimens of the bioplasm of mucus, of epithelium and of cartilage, which illustrate the same point. In every case the continuity of the bioplasm of the little outgrowth with the general mass within the cellulose wall can be well seen. In Fig. 8 are represented some of the smallest and simplest of the yeast particles under a magnifying power of 2,800 diameters, the $\frac{1}{50}$ of an inch object glass. Both these drawings illustrate the mode of formation of *germs* in this very simple organism, and it will be shown in another part of this work that the bioplasm of man and the higher animals gives off diverticula in much the same manner; and that these when detached become free bioplasts.

The diffluent germinal matter within the yeast cell is the material upon which alone all growth and action depend. Were it not for the bioplasm or germinal matter the cell would be lifeless and passive—incapable of "exciting fermentation" or any change whatever. Every particle of the bioplasm is living, and may, under favourable circumstances, undergo development into complete yeast cells, so that by the artificial division of one, thousands may result. And if the soft bioplasmic matter which can be expressed from the yeast cell be placed under favourable conditions, every particle of it may germinate. This matter alone furnishes the germs, it alone grows and appropriates the nutrient material, in short it alone manifests phenomena peculiar to living things.

The little buds or gemmules above referred to, detached from the parent mass, and capable of independent existence, are, many of them, much less than the $\frac{1}{100,000}$th of an inch in diameter. But each is living and will grow, under favourable circumstances, into a body like the parent cell, giving origin in its turn to countless descendants. These very minute particles divide and subdivide independently, producing still more minute particles, capable of growth and division like themselves, not one of which, however, may be developed into an ordinary yeast cell like those represented in Fig. 6, plate I. This mode of multiplication may go on for a long period, perhaps for an indefinite time, if certain conditions persist. But if any one of these excessively minute particles falls into a medium containing suitable pabulum, it will appropriate it, and soon pass on to a higher stage of development. In this case branches may be formed, as represented in Fig. 11, plate II., and more advanced in Fig. 12. See also Plate III., fig. 22. From them may proceed stems which grow upwards into the air,

GERMS.—FUNGI.

Fig. 9.

Old spores of fungi, in which the formed material has become very thick. At *a* the bioplasm is passing through pores. Its further growth is given in Fig. 10. × 1800 pp. 12, 16.

Fig. 10.

Germs of fungi, showing the relation of the bioplasm: the formed material and the production of the latter from the former.
× 1800. p. 2.

Fig. 11.

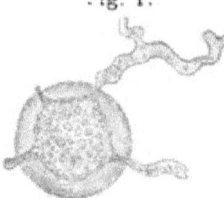

Passage of the passive forest...
into the mass of which...
contain, bioplasm and...
fungi × ...

Fig. 12.

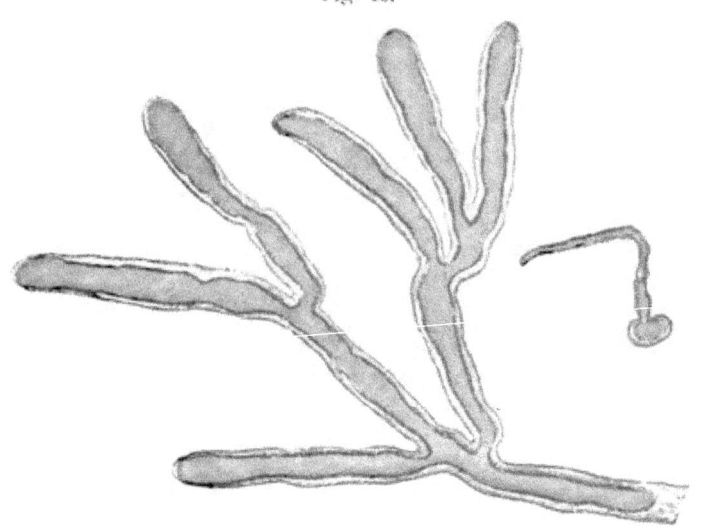

The extremities of a branching stem of a rapidly-growing fungus from jam. At the spot where the cellulose wall is only just forming, it is so thin as to be hardly demonstrable. The germinal matter in this situation is abundant, and it is here principally that growth takes place. × 215.
a, a spore which has just commenced to sprout. × 500. 1869 pp. 12, 16.

Fig. 13.

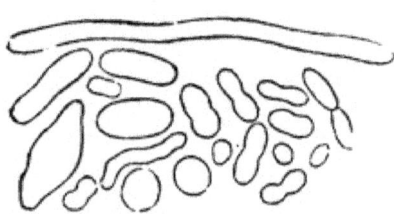

Most minute bacteria growing and multiplying. Shown in outline. × 2800. pp. 21, 34.

Fig. 14.

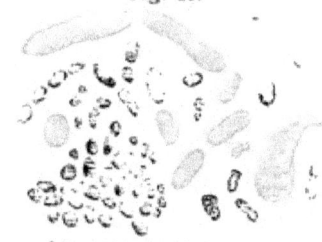

Germs of bacteria, growing and multiplying rapidly × 1800. 1885. pp. 21, 34.

$\frac{1}{1800}$ of an inch ——— × 215 linear.
$\frac{1}{15000}$,, ,, ——— × 1800.

and bear upon their summits heads in which spores are formed, these last being so well protected from the influence of destructive agents, that the germinal matter within can retain its vitality for a great length of time. Fig. 10, plate II.

GERMS IN THE AIR.

The spores just referred to are so light as to be easily supported in the atmosphere, and they may be carried a long distance by currents of air. The infinitesimal germs before adverted to are of course transported still more readily. Some of these are represented in Fig. 1, plate I., and some of the most minute that can be discovered with the highest powers of the microscope, in Figs. 1 to 5, and in Figs. 13 and 14, plate II. Bodies weighing 100 times as much as these can be supported in air. There is, therefore, no wonder that germs so very small and light are almost constantly present in our atmosphere. Myriads no doubt perish for one that falls in a spot where it meets with suitable food and other advantageous conditions. It is not to be wondered at that such minute germs as these should exist almost everywhere, floating in the atmosphere, deposited upon everything, and ready to undergo development wherever the conditions may be favourable. And the vitality of the bioplasm of lowly organisms of the kind under consideration is so great, that protected as it is by its external envelope of formed material, it is able to resist effectually and for a long time the disturbing influence of adverse external conditions. Should these lead to the death of a great part of the living matter or bioplasm, it must be remembered that by this very change, the thickness of the external envelope is increased, and thus the speck of living matter which remains, becomes still more effectually protected than it was at first (see Figs. 9 and 10, plate II.).

Protected in this way, multitudes of such germs escape destruction. Provided only a speck of living matter remains, and resists the influence of adverse conditions, it will increase under favourable circumstances. Particles will make their way through pores in the envelope, and coming into contact with the pabulum outside will soon increase and develope minute germs or branching stems as represented in Figs. 11 and 12.

Of the Detection of Vegetable Germs in the Air.—Vegetable germs were detected in the air more then twenty years ago by a number of observers, and by many different plans of procedure. One method was to cause air to be drawn through a glass vessel, the outside of which was cooled by ice. Upon collecting the water which condensed upon the interior in a vessel placed beneath for its reception, and examining under the microscope the slight sediment which subsided, the germs were discovered. Another plan was to cause air to be projected against glass plates, the surface of which had been wetted with weak glycerine, water, or some other substance, to which the particles would adhere; or merely to hang up such plates in the air for a time. Some observers employed paper moistened with the same fluids. Pasteur investigated the matter very carefully, and showed that the number of germs existing in the air varied much in different places and at different heights. He also found that the air which had been undisturbed for some time in cellars was almost destitute of germs. Mr. Crookes followed a plan which had been previously adopted by Schrœder, and separated germs from the air by filtering it through pure cotton wool. He also collected the germs in tubes, and upon glass slides moistened with glycerine (Cattle Plague Report, 1866). It would be generally concluded that many of the germs suspended in air and capable of being wafted long distances by currents, would subside if the air became perfectly still, just as the dust in our rooms falls upon the floor or is deposited upon shelves and other projections. Dr. Tyndall has, however, considered it necessary to *demonstrate* the fact that tiny particles of dust really do fall down to the ground if the air in which they are suspended be quiescent. By throwing a ray of very bright light upon air contained in a closed chamber of glass 3 ft. by 2 ft. 6 inches, and 5 ft. 10 inches high, tapering to a truncated cone at the top, he was able to prove not only that particles of dust which float in the air for a time after it has been disturbed gradually subside, but that these were deposited upon the sides of the chamber as well as upon the floor. " The chamber was examined almost daily ; a perceptible diminution of the floating matter being noticed on each occasion. At the end of a week the chamber was optically empty, exhibiting no trace of matter competent to scatter the light." This experiment is calculated to convince any one who doubts the fact that particles of *lifeless* matter, even when very minute, really do tend to fall through air, until they reach some resting place which prevents them from falling still lower, and the great discovery is announced that ponderable bodies are ponderable. But the demonstration determines nothing whatever with regard to *living* particles, not even their presence.

Dr. Tyndall also makes a prediction concerning the discovery of the cause of the impotence of certain air as a " generator of life," as if *air*, which is lifeless, could under any circumstances *generate* that matter which lives. Has any one been educated into a belief in the *life-generating* properties of the air ?

The nature of the materials which make up the dust of our air is a subject which has often engaged the attention of microscopists, who have long been familiar with the fact that multitudes of organic particles derived from various sources are floating in the more or less disturbed air of our rooms. Many of these fall into our microscopic specimens while they are being mounted, and in spite of all our care, and the use of glass shades and other means, we often preserve portions of hair, feathers, scales of the wings of insects, and a number of other foreign bodies which we would gladly exclude from our preparations. The characters of these organic particles are but too well known to us, and probably hundreds of microscopists have many times examined the dust which commonly collects upon shelves and little projections from the walls of our rooms, for the very purpose of demonstrating the many different kinds of organic fragments of which it is in great part composed. Memoirs have been written upon the extraneous matters which fall into urine, sputum, and other secretions which it is the business of the physician to examine. These extraneous particles which have been deposited from the dust suspended in the air of the rooms have given rise to great confusion, and some of them have been mistaken for bodies derived from the organism of man. No wonder, therefore, that much attention should have been given to the examination of dust, and in order to prevent mistakes figures of some of the most important constituents of dust have been given.* Particles of hair and wool of various kinds, filaments of cotton and silk, portions of insects, especially the scales of the common clothes-moth, starch granules, pollen grains, fragments of wood, and animal and vegetable germs, are among

* See for example "How to Work with the Microscope," 4th edition, p. 195, Plate XLIV.

the many organic constituents of dust which are familiar to microscopical observers. Pouchet found in the dust of the air (*Comptes Rendus*, March 21st, 1859), "the detritus of the mineral crust of the earth, *animal and vegetable particles*, and the minutely divided *débris* of the various articles employed in our wants." Mr. Samuelson, many years before 1863, obtained living germs from dust taken from the window panes, and from other common-place localities, and gave numerous figures of the different forms he discovered ("On the Source of Living Organisms," Quarterly Journal of Science, Vol. I., p. 598). Many other observers have also examined dust of various kinds with great care, and have described the organic particles existing in it in great numbers.

Yet in spite of the numerous observations which have been made and published upon this subject, we find Dr. Tyndall teaching the public in a lecture on "*Dust and Haze*," given at the Royal Institution on Feb. 18th, 1870, and afterwards published in several newspapers, that he had only just discovered that the dust of our air contained *organic particles*. His remarks were afterwards more widely diffused in several journals under the title "Dust and Disease," no connection whatever having been shown by the lecturer to obtain between disease and dust or between either and germs. Dust may certainly be perfectly harmless.*

* Witness the cavalcade of picturesque stalwart women crossing Hyde Park every evening on their way from the dust-heaps at Paddington, to

On the other hand, is it impossible that disease germs may exist free from dust?

Dr. Tyndall had thought "with the rest of the world, that the dust of our air was in great part inorganic and non-combustible." But from the "rest of the world" must be excluded the majority of those who have used a microscope or are acquainted with the use of this instrument. Ought a lecturer endeavour to excite the astonishment of his audience by trying to convince them that until lately he was quite unacquainted with some things very generally known, and when his hearers have become sufficiently interested in this want of information on his part, relate as some new discovery what he has happily to introduce to their notice? Dr. Tyndall has discovered that the little particles of cotton and hair and wool and feathers and other organic substances which exist in the air, can be destroyed by a red heat and converted into smoke, which would not be the case had they consisted of inorganic matter as he supposed before he tried the experiment he proceeds to repeat.*

their homes at Westminster. Whatever may be said against them, they are evidently persons in rude health and more affected by the passive organic and inorganic constituents of the dust than by its active disease-producing germs.

* Dr. Tyndall, after the publication of his lecture, announced in the "Times," that Dr. Percy had discovered that the dust upon the walls of the British Museum contained 50 per cent. of inorganic matter —this, in support of Dr. Tyndall's belief, that the dust of our air was in great part inorganic and non-combustible. But Dr. Tyndall says nothing about the nature of the remaining 50 per cent. of this dust.

But more than this, Dr. Tyndall professes to be able to bring "the air of the highest Alps into the chamber of the invalid." This grand result he proposes to achieve by causing the dusty air to traverse cotton wool, by which operation the dust which had been demonstrated together with the germs supposed to be mixed with it are filtered off. Even Dr. Tyndall will scarcely be inclined to deny, after a little quiet reflection, that the promise to bring Alpine air into the London sick rooms, may appear to unromantic people who actually attend upon invalids more like the result of emotional excitement than a conclusion deduced from any exact methods of observation or experiment.

By the physical method of examination, particles of wool and cotton and hair, scales and other particles from insects, and starch and soot, and all the other constituents of dust, alive and dead, organic and inorganic, are illuminated so as to form one confused ray, which can be seen at a great distance; but, it need scarcely be said, the brightest light the physicist can cause to beat upon them fails to reveal the nature of the several dust particles, or enable anyone to distinguish the living particles from the lifeless *debris;* or the virulent disease germs, should there be any, from the harmless dust. Now, instead of burning all the organic matter together, the living and the lifeless with the non-living, including that which exhibits form and structure, and that which is formless and structureless, and demonstrating the smoke resulting from the destruction of all, any one with the aid of a microscope in less time, with less trouble, without complicated apparatus, might have quickly demonstrated, and with the greatest precision identified the various kinds of organic particles present in that particular specimen of dust, and he could have shown the particles themselves instead of the smoke which resulted from their combustion and destruction. In short, the method of burning dust proves only what can be proved quite as positively by much simpler means, and proves absolutely nothing regarding the existence of disease or other germs in the air—a subject which has been successfully studied during the last ten years or more by other methods of inquiry. It is difficult to imagine anything further removed from fact than the statement that the dust of our air consists of disease germs. No one would dare to make such an assertion in plain words. It would be nonsense. And yet remarks have been made which have undoubtedly led the public to infer that such a conclusion was implied, or that it was desired that such an inference should be drawn. It is, however, still more astonishing that writers in some of our leading journals should be so misled as to give, without comment, an interpretation of an author's views which amounts to absurdity. Thus the "Academy" (April 9th, 1870, p. 185), in referring to Dr. Tyndall's lecture on Dust and Disease remarks, " The floating dust-like matter revealed in the air by a sunbeam is *organic, probably germs of animal and vegetable life.*"

In a subsequent letter to the " Times " (April 21st), Dr. Tyndall says that the organic matter of London air is not all "*germinal matter!*" The lecturer has spoken of germs, animalcules and germinal matter, but he does not explain whether these terms are used by him for the same thing or for different things. He speaks of the " morass of wordy discussion and contradictory argument," in which he says, without identifying the offender, " the germ theory of disease and the question of spontaneous generation are entangled," and with surpassing simplicity he assures the editor of the " Times " that he is " unpledged to more than a clear statement of the germ theory," and refers to the exactness of physics and chemistry, and to the methods by which he proposes to detach " from the domain of vagueness and uncertainty each successive fragment of demonstrated truth ! "

In common fairness, Dr. Tyndall should have studied the matter a little before he gave his first lecture, instead of in his last characteristically excusing himself on the ground of " tenderness for the public,"* for not alluding to the researches of others, and for with-

* The considerate tenderness which, as a rule, is reserved exclusively for the public, is, it must be admitted, on one occasion extended towards an individual. Dr. Tyndall did indeed admit that he could not read without "sympathetic emotion" the papers of an observer, whom, however, a stern sense of justice compels him to regard as "a man of strong imagination"—as one "who may occasionally take a

holding from the press statements concerning these which he had "twice or thrice written down." Had he done so, there would be less reason to condemn or to laugh at the following paragraph with which he concludes his last letter to the Editor of the "Times," and retires for the present from the consideration of dust, disease, and germs. "For a long time to come I shall be unable to devote any attention to this subject, and this has caused me to write at what I fear you will consider inexcusable length: were the question of less practical interest to humanity, I should not have troubled you either with this or with any former communication."*

It would indeed be difficult to point out a series of conclusions less justified by the experiments.

flight beyond his facts!" but he tenderly observed, "as long as the heat (dynamic heat of heart by which alone the solid inertia of the freeborn Briton is to be overcome) is employed to warm up the truth without singeing it overmuch; as long as this enthusiasm can overmatch its mistakes by unequivocal examples of success, so long am I disposed to give it a fair field to work in."—"Times," February 19th, 1870. Such comments are very curious, but they are not in good taste, and are altogether uncalled for and out of place. Dr. Tyndall had no right to speak in the way he has spoken of such a man as William Budd, of Bristol.

* Dr. Bastian has expressed his opinion of Dr. Tyndall's still more recent observations upon the influence of germs and animalcules upon disease as follows:—"The question is, however, one of so complicated a nature, that little save amazement will be excited in the minds of those conversant with all the difficulties of the problem, that Professor Tyndall should place so much reliance upon indirect evidence towards its solution, and should step forward on the strength of this, with the view of establishing a doubtful theory of disease, to which he, by his own confession, has so recently become a convert."—"Times," April 13th, 1870.

It is difficult to see in what way Dr. Tyndall's experiments illustrate or affect any sort of germ theory. They could all have been made without a single reference to germs of any kind or to a germ theory or to disease. On the other hand, these last subjects might have been rendered more intelligible to the unlearned if their attention had not been diverted by the brilliant illuminations and combustions.

The *method* is defective. The Professor first points out that he supposed that dust consisted of inorganic particles. He discovers to his surprise that it contained organic matters. Next he seems to wish that his audience should regard these organic matters as germs or animalcules, or at any rate look upon some of the particles of dust as composed of "germinal matter." But Dr. Tyndall did not demonstrate what the organic particles in air were, nor did he prove that dust contained anything whatever that would give rise to disease. The particles might have consisted entirely of harmless germs, or of disease germs or of animalcules, or there might have been a few of these bodies present, or there might not have been a vestige of any of them, and Dr. Tyndall would not have discovered the difference by the method of investigation he employed, of the advantages of which he speaks so confidently. By his discoveries, surmises, assertions, and predictions published in the "Times" and other journals, people have been led to suppose that dust consists of germs, and that air teems with animalcules and disease-producing particles, which is not really the case.

There is now a strong feeling in favour of scientific teaching. And branches of science, particularly those capable of illustration by experiment, are deservedly popular. But if the feeling in favour of scientific education is to be lasting, and not a mere fashion, and if it is desirable that the public should have any respect for science, her exponents must not put out what they have to teach in a sensational form. Undoubtedly the public may with good reason find fault with many of us for being slow and dreary, dry and uninteresting, and for presenting our lessons in a hard, unpalatable, not easily digestible form. But in endeavouring to escape these faults, it is very undesirable that any tendency towards the gushing and hysterical or rhapsodical, should be permitted. Our object it must be distinctly understood is to teach, and not to excite or surprise or amuse. There are theatres, and in great number, established for the very purpose of affording amusement where we may enjoy excitement and wonder and surprise to our heart's content, but the scientific workroom is built for instruction and for real earnest work.

The Characters of the most minute Vegetable Germs.—The very minute vegetable organisms which may be obtained from the air, and which are developed in infusions of animal and vegetable matter, are for the most part of an oblong oval form, frequently exhibiting a constriction which corresponds to the point of division (see Fig. 20, plate III.). Some are, however, much more elongated than others. Compare Figs. 15, 16, 21, plate III., with Figs. 1, 2, 3, 4, plate I.; but whether these represent different species, or are merely variations due to the circumstances under which they have been developed, is not known. Sometimes a very elongated form is found amongst numerous short ones, Plate II., fig. 13.

The general appearance of minute germs of fungi multiplying rapidly in fluid favourable to their development, is represented in Plate I., fig. 1, under a power magnifying only 200 diameters. The most minute germs (bacteria) visible under the higher powers are seen in Figs. 2, 3, and 4, which are magnified respectively 1,800, 3,000, and 5,000 diameters linear. In Fig. 5 the appearance of very minute particles of bioplasm, which have been well stained with carmine and examined under the $\frac{1}{70}$, which magnifies 2,800 diameters, is represented.

Some notion of the manner in which the most minute germs multiply, may be formed if Figs. 13 and 14, plate II., opp. p. 20, be carefully examined. Whether there is any actual firm membrane around the minute particles of bioplasm represented in Fig. 14 is very doubtful. It is more probable that each little particle of bioplasm is embedded in a soft and semi-fluid formed material which has been produced by it. This, under certain circumstances, may become condensed, and thus an envelope or protecting covering may be formed. The particles in Fig. 13 are only represented in outline.

The minute germs developed in infusions in closed vessels are represented in Figs. 17 to 20, plate III. These are referred to in page 49. But the somewhat definite characters manifested by these particles last mentioned, are not exhibited by the most minute germs discovered by the aid of the highest powers. These, like the most minute particles of other kinds of bioplasm, always appear as little specks of a rounded form. There is no possibility of identifying the different kinds of bioplasmic matter under the microscope. The most minute living particles of a vegetable organism exactly resemble those of an animalcule or those which may become developed into beings still higher in the scale; and in another part of this work it will be shown that these cannot be distinguished from particles of bioplasm derived from the living *mucus*, or *pus*, or *white blood*-corpuscles of man himself. Every kind of bioplasm at this stage of its being exhibits, as far as has yet been ascertained, precisely the same characters.

Of Germs of different kinds of Vegetable Organisms.— Of microscopic fungi and algæ there are many different kinds, which grow and multiply under very different external conditions, and live upon different kinds of food. Thus there are germs of numerous different species diffused through the air and wafted long distances at different seasons. Some flourish at a temperature which would be fatal to others; some live upon vegetable, some upon animal, matters. Some require solid substance upon and into which they may grow, while others seem to obtain from the atmosphere alone all the materials required for their growth and development. Some enjoy light, while others vegetate freely in darkness. Many of these vegetable germs are almost constantly diffused everywhere in the atmosphere, ready to increase a millionfold in a few hours whenever circumstances should be favourable. But all increase and grow in the same manner; all consist of the growing, living, active, moving *bioplasm*, or *germinal matter*, and a certain proportion of the passive, lifeless *formed material* around it, which has been already referred to.

It is supposed that germs of different species of vegetable organisms give rise to the phenomena in the system invaded, which are characteristic of the several contagious diseases, and by which they are recognized and distinguished from one another. In the germ stage, however, there are no characters which would enable us to determine the source of the germ; and whatever differences may exist in the fully developed state, at an early period of existence the embryonic living particles are alike.

It is not my purpose to direct attention to the various species of microscopic fungi which are known, or to discuss the vexed question concerning species and variety, or to indicate the variety of appearances which may be assumed by one species, and which seem to be determined rather by varying external conditions and food than to be due to inherent specific powers. These questions are interesting and important enough, but I must now pass on to consider several points more intimately connected with the question concerning the origin of vegetable germs, and their supposed influence in causing disease.

Of the Origin of Vegetable Germs.—Several scientific authorities of high repute have of late summed up very distinctly in favour of the doctrine of the formation of living beings *directly* out of lifeless matter, without the instrumentality of pre-existing living matter. On the other hand, there are men well qualified to form an opinion who consider that the advanced minds of the present day have been led to strain facts more than was right, in order to make it appear that spontaneous generation, as well as some other doctrines correlated to this, ought to be accepted. The so-called "tendency of thought" has been adduced in favour of these views; writers of advanced articles in our magazines, distinguished for the brilliancy of their speculations, have written up the doctrine; and there is reason to think that no inconsiderable number of readers is desirous of being told, and is really anxious to believe, that living things may come direct from lifeless matter, and that force may build up structures and form organs without the assistance of intelligence, or the intervention of creative power or other mysterious agency; and that matter requires but to be exposed to the influence of certain conditions in order to assume the living state. Nevertheless, every one admits that, in all the instances he knows about, the living being did, without doubt, come from a pre-existing living being. But, he proceeds to argue, because a *highly complex creature* cannot be formed direct from the dust of the ground, does it therefore follow that *all the simple living forms* modern research has brought under our observation have come from pre-existing simple creatures like themselves? Why, he asks, are we to assume that a simple structureless mass of jelly *must* come from a pre-existing mass of jelly? There must, he urges further, have been, at least in the beginning, a beginning of life. The living, at some time or other, did spring direct from the lifeless. If, then, it is admitted that this has happened once, is it unreasonable, he might ask, to conclude that it happened more than once, nay, many times, nay, he might remark, may we not feel sure that it has happened lately, and is going on daily and hourly?

It is very generally admitted that ever since life first appeared on our earth uninterrupted development has proceeded, and it is maintained by many that the evidence in favour of the view that the higher forms have been derived by descent from lower ones, is almost conclusive; and, it might be said by those who accept the doctrine, since " spontaneous generation is still in operation, the lineal descendants of the simple beings which are now being evolved in vast numbers direct from the non-living, will, after the lapse of ages, during which progressive change shall never cease, become the parents of highly developed organisms totally distinct from any which have yet existed, and of which we, with our imperfect knowledge of the properties of the particles of non-living matter, cannot form even the faintest conception. But if our knowledge was sufficient, we should be able to determine now the specific characters of the creatures that are to be in the ages yet to come." May we not argue back with equal justice that the producer is greater than the thing produced, and that, therefore, we ought to go at once to the fountain head for life, which on this theory would be the simplest non-living matter? Further, it has been argued, " since we can trace a certain gradational relationship between the higher and the lower forms of living beings, we may consider it proved that the non-living is related to the lowest, simplest living in the same sort of way." The new philosophy, after affirming that the higher life passes by imperceptible gradations into the lower life, enquires, can it be possible that any one with intelligence should doubt for one moment that the lowest life passes gradually into the non-living? And modern philosophy boasts of her exactness, and professes to accept nothing that cannot be proved by observation and experiment. But her disciples occasionally forget the iron rules they have laid down for others, and here, as in some other instances, after asserting that as the first position is proved, the second must be true, they just suggest that if not *proved* to be true, it is at any rate *capable of proof*, or if not actually provable just at this present time, is sure to be *proved to demonstration before very long!*

Any objection to such remarks as these, which are pronounced to be in harmony with the tendency of modern thought, are accounted frivolous. But, fortunately, or unfortunately, according to the stand-point taken, every one does not feel able to accept these arguments, and in spite of being considered foolish enough to attempt to oppose the whole tendency of modern thought, which is undoubtedly difficult, and perhaps impossible, I must, nevertheless, venture to remark that the doctrine of heterogenesis has not yet been proved to be true as respects one single living organism, and it is unfortunately the case that some of the "facts" which have been adduced in its favour are not facts at all, while of the "facts" some have been misstated and many misinterpreted.

He who does not accept the doctrine of the hour is, in these days, in danger of being denounced as a bigot, and stigmatized as *orthodox* because his rejection of the physical and sensational is held to prove him to be a bigoted believer in every antiquated doctrine that has ever been demolished by the bright light of research in the time that has passed. He is also likely to be reproached as a heretic because he refuses to bow down to the phantom called the tendency of modern thought. But unreasoning credulity is not peculiar to old beliefs, neither is persecution. If the bigot of former days set physics at defiance, does not many a modern philosopher unquestionably attribute to physics phenomena which are altogether beyond the range of merely physical law? It is even doubtful if the unreasoning faith of the skilled scientific does not sometimes exceed the vulgar belief of the poor ignorant bigot. If the bigot is to be accused of believing in spite of reason, the sceptic of modern times sometimes exposes himself to the charge of justifying his disbelief by argument which has been proved to be false, as well as by advancing as a fact what is really but an assertion in a fact form. If the bigot may be laughed at for his belief in the unseen and unknowable, how is the modern enthusiastic believer in the omnipotence of the material to escape ridicule?

However absurd, and against all the accumulated evidence of observation, it may be to believe in the immediate creation of any particular species of complex plant or animal, it is at any rate equally absurd, and also *quite contrary to any evidence yet obtained*, to maintain that living forms result from the direct combination of particles of inanimate matter. And whatever may be said in favour of the uninterrupted continuity of life, and of the gradual alteration of living forms from age to age—all that has been proved in connection with the growth and development of every class of living beings, tells against the doctrine of heterogenesis. And many who have written in its favour have convicted themselves of inconsistency, unless it be consistent to believe at the same time in the law of continuity and succession, and in a law which involves discontinuity and interruption as applied to the production of living forms at this present time.

I will, however, admit that upon such a question as heterogenesis, any one acquainted with the facts and arguments on both sides, and well accustomed to the marshalling of evidence in order, would, by a judicious selection of his facts, soon adduce evidence which would convince the unlearned of the truth of that view which he chose to advocate. Just at this time many circumstances have fostered in the public mind a demand for arguments in favour of the origin of living beings from mere matter independent of a superintending will. People have been so well educated that they do not recognise the oft-repeated assertion, that living beings are being continually made out of inanimate matter without the aid of already existing living beings, as a mere dogma, in direct support of which well authenticated facts and conclusive experiments cannot be adduced. That with some heterogenesis is accepted as an article of belief is not disputed here. There is no harm in this, but that there is reasonable ground for the faith must be distinctly denied. And it is probable that the doctrine would have attracted very little attention if it had not been forced into undeserved notoriety in consequence of the support it seemed to afford to modern matter-and-force views.

There has been far too much tendency of late to decide scientific questions by a show of hands. A very little tact is required to make a number of people who know little about the matter look with favour upon a new theory, for in their enthusiastic haste to upset old creeds of which they are tired, they are sure to neglect to ascertain whether the new one they accept is really as reliable as the old one they discard But the cry " Hurrah for spontaneous generation," will not advance the cause, for happily, science is not like politics, in which people may take sides and settle things by acclamation, and action has to be determined by expediency and a number of considerations quite apart from the mere question of truth or of fact. In science, views are changed in no time, and theories most popular for a while are discarded the instant some new fact is revealed. Nothing can however retard scientific discovery more than the attempt to convert scientific deliberations into mere party questions. Science is open to all the world, and although excuses have been made for spreading inaccuracies on the ground that it was necessary to put the subject in a form to please unlearned persons—such excuses are utterly inadmissible. The unlearned public can understand any scientific question that is put before them clearly if they choose to take the trouble to do so ; and he who sacrifices accuracy to brilliancy and sensationalism offends in two directions, for he can have no respect for the intelligence of the people whom he really misleads, but is professing to teach, and he brings discredit upon science, gaining only for himself thoughtless applause. People who care very little about scientific investigation are encouraged to express themselves convinced of the truth of this or that scientific doctrine, just as vast numbers have believed in the truth of the table-turning, spirit-rapping, and other fanciful manifestations. They appeal to evidence which they assert to be convincing to their judgment, although the favourite views may be opposed to known laws, and be inconsistent with demonstrated facts.

That ninety-nine hundredths of the living beings on this globe should be derived from living beings that existed before them, while one hundredth, or one thousandth, or one ten thousandth should result directly from non-living matter, is very improbable ; but because it is affirmed that such a view is "advanced," and in harmony with the whole tendency of thought, people do not stop to consider its probability or improbability. Continuity prevails as a law, but continuity is not to be universally applicable. Experience and observation demonstrate in thousands of cases that living matter is derived from living matter, and yet we are asked to believe that in some instances living matter comes from lifeless matter, because the active and energetic minds of the day assert this. We are told authoritatively that we must believe that the non-living passes by gradations into the living, although the bodies supposed to establish the fact of these gradations exist only in the imaginations of those who make the assertion. It is unsatisfactory, if not useless, on the part of any one, however great the authority he wields, to declare that although substances in a state of transition from the non-living to the living cannot yet be produced, they *will be discovered at some future time.* Why are we to believe him? Of all prophecies the prophetic assertions of the scientific are the least worthy of belief, for the scientific spirit is utterly incompatible with the spirit of prophesy. No one can have so mistaken his calling in this world as the scientific man who ventures to prophesy, and yet the prophetic spirit seems to prevail in scientific quarters where it would be least expected. Even Professor Huxley cannot quite resist the temptation of foretelling what will be possible in a certain time. He says that he believes it possible before half a century has elapsed, that man may be able to take inorganic substances, such as carbonic acid, ammonia, water, and salines, "and be able to build them up into protein matter," and that that protein matter may "begin to live in an organic form." Of course he does not consider it necessary to give any reasons for such a strange opinion. It is supported by authority, but where are the facts which support it?

Not one of the many who have joined the ranks of the heterogenists has succeeded in giving us any conception of what, according to the doctrine he accepts, really occurs when the non-living matter becomes alive. No one who affirms spontaneous generation has ventured even to theorize upon this. Is it that the supporters of this view are at a loss to conceive what takes place, or are they afraid to commit themselves to an opinion concerning the rudimentary facts upon which the faith they profess to believe rests? What can that doctrine be worth the fundamental facts of which are not to be examined? To maintain that lifeless matter spontaneously assumes the living state, and not be able to give the faintest notion of what occurs when the change takes place, is not furthering investigation but sowing dogma. No one has any right to assert that the non-living can become living without the influence of the living, because at this present time such a change is against experience and is not conceivable. Marvellous, indeed, must be the change which occurs at the moment when the living imparts to the particles of the non-living its wonderful powers. Anyone who has seen living matter increasing, and moving, and dividing, will feel that it is useless to attempt to divine the nature of the change which takes place. How utterly impossible, therefore, must it be to conceive what may occur upon the hypothesis that the non-living becomes living without the intervention of any living matter whatever.

Fig. 15.

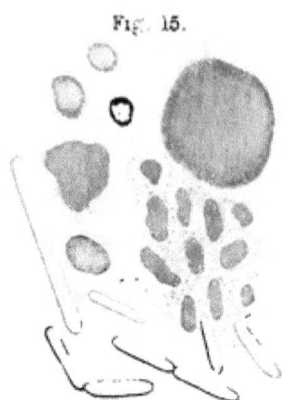

Fig. 16.

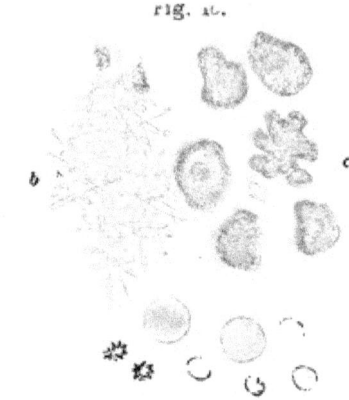

Masses of protoplasm and vegetable organisms (bacteria) in active movement from surface of villi in cattle plague. The bacteria are only shown in outline. × 1800. pp. 34, 66.

From central part of a dark red clot in the aorta—cattle plague—within twenty-four hours after death. *a*, largest red blood corpuscles and those of average size some were pale. *b*, vegetable organisms, bacteria shown in outline. *c*, white. *d*, corpuscles exhibiting active movements. × 700. pp. 3, 6.

Fig. 17.   Fig. 18.   Fig. 19.   Fig. 20.

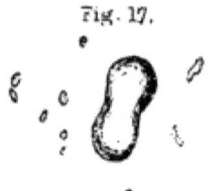 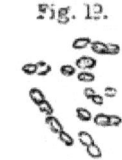 

Living organisms found in closed vessels into which fluid and organic matter had been previously introduced, due care having been taken to exclude air. The contents of the flasks had been well boiled. The large oval bodies in Figs. 17, 18, are crystals. Fig. 17 × 1700. Fig. 18 × 1700. Fig. 19 × 1700. Fig. 20 shows one of the smallest objects in Fig. 19 × 1700. pp. 34, 49.

Fig. 21.

Vegetable growths in the mucus of the gall bladder. Cattle plague. These organisms were very numerous. Below *a*, a myeloid corpuscle. × 1800. The vegetable organisms only shown in outline. p. 66.

Fig. 22.

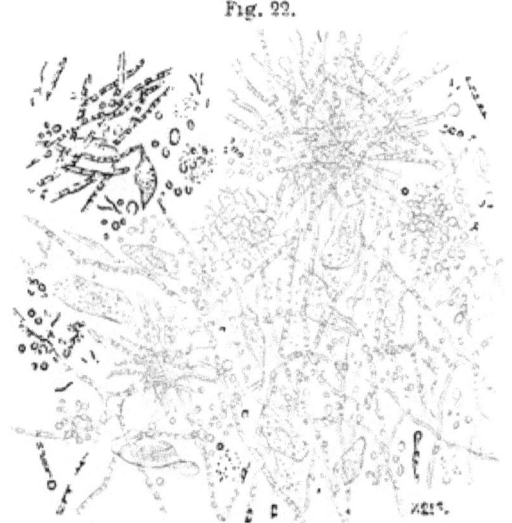

Fungi in different stages of growth in the sputum of a patient in the last stage of phthisis. Spores or germs are seen to be very numerous, and the stems have grown from these. × 215. p. 20.

1/1000 of an inch ——— × 215 linear.
  ,,   ,,   ——— × 700.
1/1000 ,,   ,,   ——— × 1800.

The results of many experiments have, however, been brought forward in favour of the doctrine of heterogenesis. Organic matter, air, and water, which have been subjected to various operations supposed to effectually destroy any living particles that may be present, have been introduced into glass vessels which have afterwards been hermetically sealed. In spite of every precaution germs have made their appearance, and it has been inferred therefore that these sprang into existence without being in any way indebted to parental organisms.

In 1864–65 I examined with Dr. Child the contents of several hermetically sealed glass flasks, into which various vegetable infusions had been introduced. In order to prevent the entrance of germs from the air, and to destroy any germs which might exist in a living state in the matters introduced into the flasks, the following precautions were adopted by Dr. Child. ("Essays on Physiological Subjects," second edition, page 116.)

"In these experiments I have adopted some slight modifications of the apparatus used in the former ones. That now employed consists of a porcelain tube, the central part of which is fitted with roughly pounded porcelain; one end is connected with a gas-holder, and to the other the bulb is joined, which contains the substance to be experimented upon. The bulb has two narrow necks or tubes, each of which is drawn out before the experiment begins, so as to be easily sealed by the lamp ; one neck is connected with the porcelain tube, as already stated, by means of an india-rubber cork, and the other is bent down and inserted into a vessel containing sulphuric acid. The central part of the porcelain tube is heated by means of a furnace, and when it has attained a vivid red heat the bulb is joined on the end of the porcelain tube, which projects from the furnace, being made thoroughly hot immediately before the cork is inserted, the cork itself being taken out of boiling water, and the neck of the bulb being also heated with a spirit lamp before it is inserted into the cork. A stream of air is now passed through the apparatus by means of the gas holder, and bubbles through the sulphuric acid at the other end. The substance in the bulb is then boiled for ten or fifteen minutes, the lamp withdrawn and the bulb allowed to cool while the stream of air is still passing through the porcelain tube, maintained during the whole time at a vivid red heat. When the bulb is quite cool the necks are sealed by means of a lamp. The advantage gained by means of this apparatus is that there is only one joint the perfection of which in any degree affects the success of the experiment, and of that joint it is easy to make sure. The porcelain tube also, being for a considerable part of its length fitted with small fragments of porcelain, all heated up to redness, easily insures that every particle of air admitted to the bulb shall be thoroughly heated."

In many instances undoubted organisms in a living state were identified, but the number present varied greatly in different flasks. The cause of the difference was not in all instances clear. In many cases the germs were so very minute, that I am quite sure they would have completely escaped observation if an object glass magnifying upwards of 1,000 diameters linear had not been employed. These observations necessarily lead us to conclude that the failure of observers who have worked with quarters and object glasses magnifying less than 500 diameters, is easily accounted for. Even Hallier, I believe, carried on his more recent observations with very low powers, and I believe the observations of both Pouchet and Pasteur are open to objections upon the same grounds.

Some of the organisms discovered in Dr. Child's infusions are represented in Figs. 17 to 20, plate III, from drawings made by myself. The large dumb-bell shaped bodies represented in Figs. 17, 18, are not organisms, but crystals. They could be readily distinguished from the living forms by their high refractive power, larger size, and absence of any movement.

It was supposed that boiling was fatal to all living things; then it was proved by experiment that some living things, under certain circumstances, did live in spite of being subjected to a temperature even above that of boiling water. But was it therefore necessary to assert authoritatively that no living organisms could live at a temperature a certain number of degrees above that of boiling water, and that in cases in which any living forms are found in fluids in closed vessels that have been exposed to that temperature, they are formed *de novo*? What is there to prevent us from coming to the conclusion supported by so many positive general facts in nature which are well known, that the living forms discovered did spring from living matter which resisted the high temperature to which they had been exposed?

Moreover, in many of the experiments it does not appear that *every part* of the apparatus had been subjected to the high temperature. If the smallest portion were left above the bath in which the closed vessel was immersed, a few living germs might have escaped the destructive action, and from these might have been developed those which were subsequently detected and supposed to have arisen in a new way.

Dr. Charlton Bastian exposed fluids to a temperature varying from 148 deg. C. to 152 deg. C. (298·4 to 305·6 Fahrenheit) for four hours, and yet in the course of a few weeks living organisms were developed ("Times," April 13th, 1870). But even this striking fact proves nothing concerning the actual origin of the living forms, and it is more in accordance with the results of observation and experiment, to conclude that living forms might live though exposed under certain conditions to a temperature even of 350° Fahrenheit, than it would be to infer that the living bodies present originated spontaneously in a fluid after it had been exposed to this high temperature. In every instance in which living forms have been attributed to spontaneous generation, the possibility of their origin from germs cannot be denied or disproved.

We have yet very much to learn concerning the influence both of high and low temperatures upon the minute particles of bioplasm constituting the germs of the lowest forms of life. And there is no doubt that the effect of the same degree of temperature would be different at different phases of the life of each species of fungus or low organism, and at different periods of the year. The effect would also vary according as the organisms were exposed to sudden great alterations of temperature, or submitted to intense cold or heat by slow and gradual changes; and even in man and the higher animals it is remarkable what great degrees of heat and cold can be borne if only the change be gradual. Some of the lower forms of life are habitually exposed to a temperature of 32, and would probably bear a very much lower temperature without being destroyed. These creatures, it must be remembered, are not merely *exposed* externally to this temperature like many vertebrata which have the power of developing heat within themselves, and whose temperature does not therefore vary with that of the surrounding medium, but they suffer every change which affects the medium in which they are placed, for their means of evolving internal heat are so slight and imperfect, that these may be left out of consideration altogether. Their bioplasm or living matter is adapted to live and grow at very low temperatures. Some organisms which do not grow and flourish at a temperature much below 50°, are nevertheless capable of bearing a low temperature, and may even live for a length of time imprisoned in solid ice. Whether the bioplasm of their organism is actually *frozen* is very doubtful. It is more likely that the bioplasm resists for a long time the process of congelation, and it seems to me probable that the motion which there is reason to think continues during life, prevents the living matter from freezing. Death most likely occurs before congelation takes place, but when once the living matter has actually become ice, its life is for ever destroyed, and it is incapable of being revivified or revitalized. It can never live or move again.

With regard to the power possessed by certain living organisms of resisting the destructive influence of a high temperature, it must be remarked that of certain of the forms discovered in the closed vessels (see p. 47), little is yet known. Many have been passed over by highly distinguished observers, and it is even probable that some have altogether escaped notice up to this very time. Of the very minute organisms in question, some may be able to resist the degree of heat to which they are exposed in the course of the experiment without being killed—nay, there *may* be several forms of organisms extremely minute which are at present undiscovered, but which among other characteristics possess the remarkable property of resisting the destructive influence of a temperature of 300° or 350°. From these many might afterwards grow.

In many instances in which the absence of living germs would have been inferred, minute organisms, invisible by the aid of the magnifying powers usually employed, have been discovered, and in considerable number. And in some cases in which it has been stated that living organisms were not present, there is reason to think many might have been detected, had greater care in the examination been exercised, and higher magnifying powers employed. But though this be admitted, the fact does not in the slightest degree strengthen the case for the heterogenists. They have to prove that living forms appear under conditions which not only absolutely preclude the possibility of the entrance of living germs from without, but which ensure the death of every living form that may have been present in the substances used for experiment. The position is indeed a difficult one, for the more investigation advances, and the more we learn concerning the minute living germs which exist in such wonderful profusion, the more do we hesitate to place perfect confidence in the means employed for the destruction of those that were present before the experiment commenced, and for excluding the entrance of living germs into the closed vessels in which the actual new generation of living organisms has been held to occur. Bearing in mind that living particles far more minute than could have been seen by the magnifying powers employed by Pasteur, Pouchet, and others, undoubtedly have been detected in the closed vessels, as in the observations of Dr. Child, for example, in which I had the advantage of assisting, and that germs are not invariably destroyed by boiling,—I would ask, is it not more reasonable to conclude that the living forms discovered were derived from pre-existing germs which obtained access to the fluids in consequence of the arrangements made to exclude them not being quite perfect, than that they had been formed anew from lifeless matter in the image of those very forms which have been unquestionably developed from predecessors like themselves? If the view of their formation *direct* from the non-living be accepted, how are we to account for their exact resemblance in form and actions to beings familiar to us, whose parentage is known; and for the fact that further growth and multiplication proceed precisely as those operations occur in germs derived from parental organisms?

But it has been suggested that although perfect living forms may not be developed spontaneously, perhaps a form from which these may soon be evolved, results in this manner. To argue from facts revealed in the course of observations made with magnifying powers so moderate as a quarter or even an eighth (two to four hundred diameters) that *eggs* are produced spontaneously is surely, in the present state of knowledge, very hazardous, if not altogether unjustifiable. Of the supposed *lifeless* particles, by the aggregation of which the eggs are said to be formed, little can be learnt from observations with powers below a twelfth, because higher objectives would have resolved these supposed lifeless atoms into something very different, and perhaps have proved, that so far from being non-living particles, they were really living organisms which had been living for some time, and were at the time of observation at any rate far enough from the inorganic. Moreover, those who have advanced this theory, and those who have given in their adhesion to it, have not intimated how we may ascertain *when* the aggregation of lifeless particles assumes the living condition; and they have left us completely in the dark as to *what* occurs when the marvellous change in question takes place. What a wonderful disturbance must occur at the instant of animation! What a violent dislocation of elements which were combined as compounds, and what re-arrangement must take place when the inanimate collection of molecules starts into vitality! What sort of force effects the change, and whence arises the destructive force and the constructive power? Not a word of explanation on all this, and yet are we expected to accept as a fact proved, the formation of spontaneous eggs!

But the lifeless "proliferous disk," from which the living eggs are supposed to emanate so curiously, when carefully examined, is at once resolved into countless millions of separate living particles, every one of which lives and grows, and must be regarded as a distinct germ. Every one of these millions may give origin to successors, each little bioplast being a distinct being with all the attributes of an independent living organism, and containing within itself all the marvellous self-propagating powers of a living germ. I feel sure that anyone who patiently studies the simplest forms of life under the highest powers of the microscope, will utterly reject the so-called *observations* which are adduced in support of the formation of "spontaneous eggs" by the aggregation and coalescence of lifeless particles. In the ten thousandth part of such a proliferous mass are living germs enough to produce by simple division in the course of a few hours countless multitudes of living forms.

Living particles, far more minute than the life-constructing, non-living particles, have been seen and studied, and they have been observed to increase and multiply. But what is the nature of the mysterious operation of vivifaction which takes place at the moment of the conversion of the lifeless into living matter? Is the passage from the inanimate to the living condition sudden and abrupt, or gradual? According to Owen, this process is going on *daily* and *hourly*, so that there ought to be abundant opportunities for studying it carefully. This authority does not tell us what he means by the daily and hourly conversion of physical and chemical into vital modes of force, but surely such an investigator as Owen will not deem it right to leave this bare assertion without any further explanation. Every one interested in this wonderful problem naturally desires that he should give us some idea of the view he has formed in his own mind regarding what takes place at the moment when the mode of the force ceases to be physical and becomes vital—when the passive atoms become active organisms—when the inanimate leaves the state of lifeless rest and assumes that of living activity—when the matter acquires converting powers which it never possessed before—when, after having collected together by aggregation, the now living matter begins a new existence, and, instead of aggregating, its particles move away from one another—separate, never to join again. Any statements affirming that living particles have been *seen* to coalesce and join, under a power of less than five hundred diameters, are not to be relied on. A mistake is very easily made, and before an observation advanced in favour of such a statement can be accepted as true, it must receive confirmation ; not only on account of the errors possibly made by the observer himself, but because the conclusion is opposed to many broad facts which may be unquestionably relied upon, and particularly the fact of the formation of these same organisms by division and subdivision, which has been observed and confirmed in many cases by hundreds of competent observers. Nothing is gained by the statement being repeated over and over again that lifeless particles of matter come together and form a living thing, save that by mere iteration people who have concerned themselves little with the subject may be persuaded to assent to the view advocated, and, without examination, proclaim it to be true. But knowledge cannot be advanced by declarations and affirmations, or by the consent of numbers.

On the other hand, many microscopists have actually seen the living particles detach themselves from a pre-existing living mass, and have abundant evidence to prove that this process takes place among creatures occupying various positions in the scale of living beings as well as in many different forms of germinal matter from the highest organisms. In some of the lower creatures the process may be watched from hour to hour, as it gradually progresses towards solution of continuity, and the formation of two beings out of one is completed. I have myself frequently witnessed the sub-division of living particles from the organisms of the higher as well as from those of the lower forms of life, so minute and of such tenuity, that they could only be seen with difficulty when magnified 5,000 diameters; and there is much reason to think that even if the magnifying power could be increased to 50,000 diameters, there would still be seen only more minute living particles growing and dividing and giving rise to particles like themselves. Are we to believe, then, on the mere dictum of authority, that living germs are formed in two ways upon two distinct principles? 1. By being detached from parent living matter; and 2. By the direct combination of lifeless particles without the intervention of any pre-existing living matter at all?

It must be freely conceded that many facts are susceptible of more than one interpretation, and may be regarded as being of different import by different minds. Nay, in some instances, the very same facts have been appealed to, and not in any way unfairly, in support of opposite and conflicting doctrines. With reference to the question of spontaneous generation, I must, however, venture to remark, that to my mind the case of those who at this time hold to the doctrine of the direct origin of living beings from non-living matter appears so hopelessly opposed to facts, that I should as soon think of believing in the direct formation from lifeless matter of an oak, a butterfly, a mouse, nay, man himself, as in that of an amœba or a bacterium.

After so many failures to force people to believe that the phenomena peculiar to living beings are to be explained by physics alone, it was natural to expect that the language employed by those who still entertain such doctrines would have become more guarded, if not more exact. But, on the contrary, the more conclusively it is proved that the physical facts yet discovered are incompetent to explain vital phenomena, the more do we find vague but most positive assertion used, as if this were as convincing as the results of observation and experiment. The more desperate the case becomes the more violently do its advocates affirm that their cause is good; the more strongly do they assert they are in the right. And yet those who are determined to support the physical theory of life need not care if facts and arguments are against them, for, in order to prove their case to the satisfaction of many persons they have but to exclaim triumphantly, "In the years yet to come new facts shall be discovered which will demonstrate conclusively the truth of the physical theory of life!"

Since cosmic vapour has produced worlds, shall not air generate life? Was not the announcement made at the Royal Institution some years ago that the sun formed the heart and built the brain, and that cattle, and verdure, and lilies, were his workmanship? Why, then, shall not the air divide with the sun these marvellous powers? May we not be indebted to the sun for the formation of complex organs like the heart and brain, and may not the air generate that much simpler moving jelly-like matter of which the simpler beings are composed?

In conclusion, the following suggestions shall be offered as an unworthy contribution towards establishing an hypothesis which may illumine the path of the physicist until he arrives at the demonstration of the physical origin of life. Since it is well known that the infinitely minute particles of cosmic vapour, of which by mere aggregation worlds are formed, are diffused through space, is it not reasonable to imagine that between these, perhaps supporting them as well as separating them, is a subtle animated vapour? Heat, as is well known, causes a re-arrangement of material particles, which may be scattered or condensed according as the cosmic forces operate upon them. This scattering or condensation would occur at a different temperature in the case of different particles, according of course to the original properties of the molecules. Certain particles of the cosmic vital steam, would, at a given temperature, gyrate upwards and distribute themselves, while others *might* approach one another and form a vital crystal. This, growing by aggregation, *might* become a spontaneous ovum, the product of evolution and formifaction, containing potentially not only a fully developed organism, but whole generations of altering forms, every one of whose specific characters *might* be defined at this very time by a sufficient intelligence! The rapid increase of physical energy encourages the physicist in his attempts to forecast the future, while it enables him to secure some fragments of the real which, but for his successful efforts, would have been for ever lost in the infinite void of unfathomable nothingness.

SUPPOSED INFLUENCE OF VEGETABLE GERMS IN CAUSING DISEASE.

The manner in which they might enter the Body.—No wonder that many of the diseases of man should be attributed to microscopic fungus germs so very small that they could readily enter his organism by any of the numerous pores all over his body. Particles so minute could easily pass into his blood through the soft mucous covering of his mouth or stomach, entering these recesses with food and water. They would not insinuate themselves into the chinks between the epithelial cells of his cuticle and *move* towards the blood, as is possible in the case of bodies which possess the power of active movement, like an amœba, or a white blood corpuscle, or a pus corpuscle, but they would extend inwards, by growth and free multiplication in the very substance of the protecting epithelium. Each new particle produced would thus get nearer and nearer to the blood, which would at last be reached by the growth of particle after particle in advance. Becoming immersed in a medium adapted for their nutrition, the germs which had gained access to the blood would grow and multiply very rapidly. Countless myriads of such germs might circulate to all parts of the body. Multitudes of these becoming stationary in the capillary vessels of the cutaneous and mucous surfaces would increase there, and might give rise to the morbid phenomena which characterise fever.

Since it has been shown that living germs of entozoa a thousand times larger than these vegetable germs may traverse man's textures, and pass long distances through the tissues and organs of his body, until the particular locality suitable for their development into a higher stage of being has been reached, it is impossible to oppose to this notion of the entrance of vegetable fungus germs any serious objection, at least from this point of view. But at the same time, before anyone who is acquainted with the facts, and reflects carefully, will accept this doctrine, he will desire to be satisfied upon many points which unquestionably require elucidation. It is not sufficient to show that such particles *might* enter the body in the *manner* suggested. But it is necessary to prove not only that they really do enter, but that they give rise to the changes in the way which is suggested by the theory in question. If the vegetable germs which have been referred to are indeed the active agents, we ought to be able to demonstrate them, for there is no difficulty whatever in demonstrating closely allied organisms where they do exist, as for instance in the epithelial cells on the mucous lining of the mouth, in which millions can be seen at any time.

Of the Vegetable Germs actually discovered in the fluids and tissues of the higher Animals during Life.— In every part of the body of man and the higher animals, and probably from the earliest age, and in all stages of health, vegetable germs do exist. These germs are in a dormant or quiescent state, but may become active and undergo development during life should the conditions favourable to their increase be manifested. Indeed, if the flow of fluid which persists in the normal state in the ultimate parts of the tissues as long as life lasts be stopped, changes take place exactly resembling those which are occasioned in dead tissues removed from the body, and kept at a temperature of 100 degrees. As has been remarked, " decomposition" takes place, and, if this decomposition is not a consequence of the multiplication of the vegetable organisms, it is at any rate certain that the growth and multiplication of these bodies are constantly associated with the change in question. There cannot be a doubt that vegetable germs exist in the internal parts of the body which would grow under the circumstances supposed.

The higher life is, I think, everywhere interpenetrated as it were by the lowest life. Probably there is not a tissue in which these germs do not exist, nor is the blood of man free from them. They are found not only in the interstices of tissues, but they invade the elementary parts themselves. Multitudes infest the old epithelial cells of many of the internal surfaces, and grow and flourish in the very substance of the formed material of the cell itself. But the living germinal matter of the tissues and organs is probably

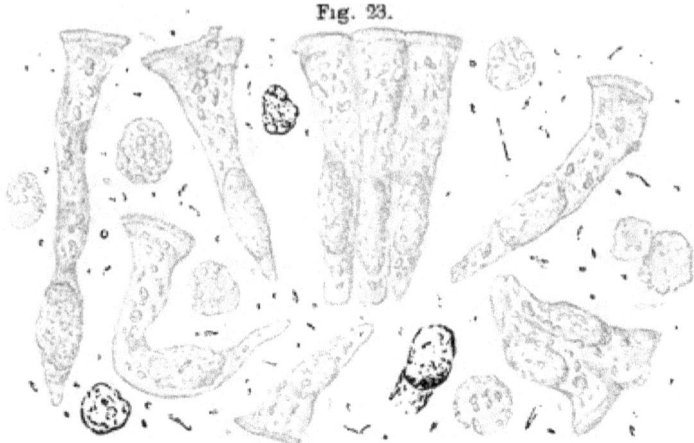

Fig. 23.

epithelium from the jejunum of a child who died of cholera. The small bodies in every part of the field and many of those in the epithelial cells themselves are bacteria and bacteria germs. These are not peculiar to cholera. They were alive when the specimen was examined. × 700. p. 63.

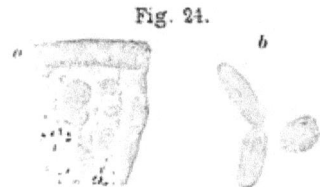

Fig. 24.

summit of one of the epithelial cells represented in fig. 23, containing germs in the interior. *a*, thick summit of the cell. *b*, free bacteria germs. × 1800. p. 66

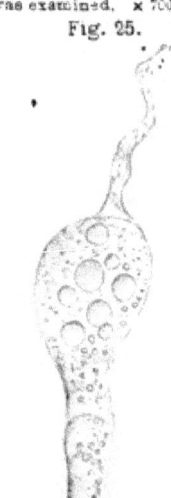

Fig. 25.

obstructed vessel, with bulgings, from the summit of a villus. Case 3. Cholera. In the interior were bacteria, oil globules, blood corpuscles, and the sporules of fungi. × 700. p. 63.

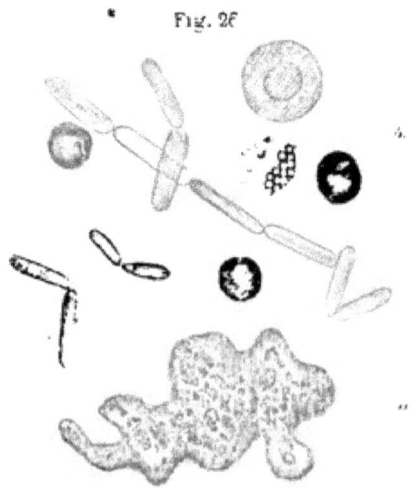

Fig. 26.

veins obtained from the hepatic vein of a cow which died of cattle plague. The round bodies *b* are growing red blood corpuscles still containing bioplasm. The large body *a* is a white blood corpuscle. The blood was quite warm when examined. × 2800. p. 66.

Fig. 27.

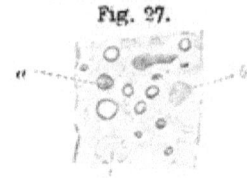

a very small portion of one of the contracted and altered capillaries from the summit of a villus. Cholera. × 2800. *a* is a small particle which somewhat resembled a sporule of a fungus. *bb*, minute particles of very smooth material (growing bioplasm). Oil globules are also seen in considerable numbers. p. 6

$\frac{1}{700}$ of an inch ——— × 700.
$\frac{1}{1800}$ " " ——— × 1800.
" " " ——— × 2800.

L. S. B., 1866. [To face p. 64.

perfectly free from vegetable germs. Some are, however, not uncommonly met with on the free surface of the germinal matter, where its death and conversion into formed material are taking place. So long as the higher living matter lives and grows, the vegetable germs are passive and dormant, but when changes occur and the normal condition departs, they become active and multiply. Millions are always present on the dorsum of the tongue and in the alimentary canal, but they remain in what may be termed a germ or embryonic state. The normal secretions poured into the alimentary canal prevent their growth, and the nourishment comes to us instead of being appropriated by them. But what happens if some of these fluids be suppressed or changed in quality? The bacteria grow and multiply, and the nourishment is no longer absorbed into our bodies. In infants a little derangement in digestion will entirely prevent the assimilation of the milk, which remains in the intestines a source of irritation, until it is expelled, serving only for the nutrition of bacteria, which are found in countless multitudes in every particle of it. If more milk be introduced it soon undergoes the same change, and the child might, perhaps, be starved by the persistent introduction of fresh food. If food is withheld for a time, the alimentary canal soon becomes emptied of its contents, and regains its natural healthy action, a process which is expedited, as is well known, by the administration of

F

some simple purgative, which excites the glands to pour out secretion, and so the passage is cleared from the stomach downwards.

In many very different forms of disease these germs of bacteria, and probably of many fungi, are to be discovered in the fluids of the body, but the evidence yet adduced does not establish any connection between the germs and the morbid process. In Plate IV. these minute organisms are represented in the contents of the alimentary canal, and in the interior of the epithelial cells of the mucous membrane of the intestine in cholera. In the contents of the blood-vessels of the same disease, and in the blood taken almost immediately after death from the vessels of animals destroyed by cattle plague and other fevers, similar bodies have been found, Plate III, figs. 15, 16, 21, though probably not of exactly the same kind in every case, Figs. 25, 26, 27.

As has been already stated, germs apparently of the same nature as those figured in Figs. 23 and 24 from cholera, are invariably to be found in the old epithelial cells of the mouth of healthy persons, and not rarely in those from many other surfaces. In the intestinal contents in various slight derangements, they are common enough, so that we cannot but conclude that their presence is due rather to alterations in the fluids *consequent upon* morbid changes, than that they are themselves the cause of the disease. They follow the morbid change instead of preceding it.

And the same observations may be made with regard to the presence of bacteria in the blood of man and animals destroyed by various diseases.

Sometimes these germs grow and multiply in a secretion not perfectly healthy, before it has left the gland follicles, and they have been detected in the milk as it issued from the breast, in the saliva, in the bile and urine, as well as in other secretions. It will no doubt be said in all these cases, " the germs have been introduced from without—they pass from the air into the orifice of the duct, and thus make their way to the gland. From this point they might readily pass into the blood." But it is more likely they are in the blood and in the tissues at all times. They are met with in the blood especially, in some instances in which there is no reason whatever for concluding they made their way into this fluid shortly before they were found. Nay, little particles may be seen in the circulating fluid which I believe to be these lowly germs, ready to grow and multiply whenever the conditions become favourable. I have seen such particles adhering to the surface of the white blood corpuscles, and also to the red blood corpuscles. In the fibrin of an aneurismal clot I have found active bacteria in vast numbers, and have observed the erosion resulting from their long-continued action so very short a time after death, that I feel quite certain they had been living upon the coagulated fibrin, and growing and multiplying during several weeks previously, and yet they had not passed into the general mass of the blood. But if this had accidentally happened, they would have been destroyed instead of multiplying, if the blood was in a healthy state. In cases in which these organisms have been discovered actively multiplying in the blood, that fluid must have already undergone serious changes, which had rendered it unfit for the nutrition of the body.

I cannot agree with those who consider that we have evidence in favour of the view that the bacteria are really the active agents in cases in which the blood has been shown to exhibit the properties of a *specific contagious virus*. The disease called *malignant pustule* has been attributed by Davaine (Comptes Rendus, 1864) to the presence of bacteria in the blood, but this observer does not prove that the bacteria were the poisonous agents, and many circumstances render it probable that other matters suspended in the blood constituted the real virus, while the vegetable organisms were but harmless concomitants. Polli, Tigri, and many others, have attributed typhoid fever and allied diseases to bacteria in the blood acting after the manner of ferments, but the objections raised to the fermentation theory have not yet been disposed of by those who advocate this doctrine. Indeed, many authorities who have attributed various phenomena to fermentation, and have spoken of the fermentation theory, have not explained what they mean by the terms they employ, and appear to have very vague notions concerning the nature of the process called "fermentation." This word is often employed very carelessly, and like "irritation," "nutritive irritability," "stimulus," and a number of other terms, is supposed to account for many phenomena, although its meaning has not been defined, and those who use it do not tell us what they mean by it.

It appears then that bacteria germs grow and multiply whenever a change takes place in the solids and fluids of the organism which develops compounds suitable for the pabulum of these living bodies. From the fact that bacteria grow and multiply not only in a few special fevers, but in a great variety of different morbid conditions, it is evident they have nothing to do with any particular form of disease. All attempts to demonstrate various constant species of bacteria, representing different contagious diseases—and many attempts have been made—have completely failed. There is greater difficulty than would appear at first in testing the matter experimentally, for it is probably impossible to introduce bacteria in quantity into the blood of a healthy animal without introducing at the same time putrescent matters which by themselves would occasion the most serious derangement. Active bacteria introduced into a healthy wound or amongst the living matter of healthy tissues, will die, although the most minute germs present, which escape death, may remain embedded in the tissue in a perfectly quiescent state.

Before the bacteria can grow and multiply, the death of the higher germinal matter must occur; as long as this lives, it, and the adjacent tissues, are freely permeated by healthy fluids, and will efficiently resist their assaults. Much as I admire the interesting observations of Mr. Lister, and firmly as I believe the facts as stated by him, I venture to doubt if the efficacy of the treatment he so ably advocates is due to the prevention of the entrance from without of these germs. There are germs out of number *within*, which would grow and multiply in the wound, however perfectly those outside were excluded, provided only the wound itself were in a state favourable to the process. Bacteria germs appear in close cavities in the substance of tissues during life and within the blood-vessels, as has been already stated. The true explanation of the undoubtedly beneficial action of the carbolic acid antiseptic treatment may be very different to the explanation offered by Mr. Lister. To me it appears much more probable that the carbolic acid acts directly upon the growth and multiplication of the bioplasm of the part; but this question shall be considered in another part of this memoir, after the mode of formation of pus has been referred to.

The virulent poison which sometimes produces such terrible results upon the healthy (?) organism in cases of dissection wounds cannot be attributed to the presence of vegetable germs, for the period of its most virulent activity is very soon after death, but before the occurrence of putrefaction and the development of bacteria.

It has been assumed that the poison in question is not developed until after death has occurred. But no one has shown that if inoculation were effected while the patient yet lived, the results would be in any way different. There is, I think, no more doubt that such poison is developed during life than that the poison of small-pox, syphilis, and many other poisons which are allied to these, and probably grow and multiply in the same manner, increase during life.

When putrefaction has actually set in, and bacteria germs are being developed in immense numbers, a punctured wound is not productive of the dire consequences which too often result if inoculation takes place within a few hours after death. In fact, the real virus loses its power when decomposition commences. Before vegetable germs appear the virus is active; soon after these have been developed it is harmless. Its power cannot, therefore, be attributed to the germs but must be due to something else which continues to live and remains active for a short time after death, and then for ever disappears. The nature, mode of origin, and multiplication of this active material will be fully discussed in the second part of this work.

Question of derivation of Fungus Germs from higher Germinal Matter of another kind.—As has been already remarked, lowly vegetable germs appear in closed cavities in the substance of dead animal and vegetable tissues. I have often seen them within vegetable cells in which not a pore could be discovered when the tissue was examined by the highest powers. I have detected them in the interior of the cells in the tissues of animals, and in the very centre of cells with walls so thick and strong that it seems almost impossible that such soft bodies could have made their way through from the surrounding medium. How are we to account for the presence of living particles in such situations? I have no doubt that ere long the theory will be advanced that the living matter of the cell, and the formed material of which it is composed, become changed, and assume the condition of bioplasm of a lower grade of organisation. Thus, it will be said, from one form of living matter a lower and more degraded form is evolved, the lowly germs springing from the living matter of the cell itself. And it might be alleged that the same forces which were once active in the cell, become active in the new organisms which grow and multiply after it has ceased to live.

Such simple germs are, as has already been stated, from time to time found in the blood of man, and to them various disturbances, ending in death, have been attributed. They have been looked upon as the germs concerned in the production of disease and in the destruction of life much higher than their own; but the matter on which they live has ceased to take part in the actions of the higher life, and instead of being decomposed into noxious gases inimical to all life, it becomes appropriated by the vegetable organisms, which grow and multiply enormously. But these having continued to increase, at length cease to multiply, and in their turn die. The products resulting from their death may serve as food for beings a little higher in the scale.

But we have now to enquire, how, if they are not actually formed there, these bacterium germs get into the interior of a perfectly closed cell. There is no real difficulty in accounting for the entrance of these germs through the cell wall; for although no pores may be visible even with the aid of the highest powers, still pores sufficiently large to permit the passage of such very minute particles as the germs are, necessarily may exist, if not in the fully-formed state of the cell, at least at an early period of its development. If we examine, under the highest powers of the microscope, fluid exudation which we know may pass through membrane, and which, when examined by ordinary means, appears perfectly clear, like water, we frequently find in it minute particles of living matter. By the $\frac{1}{50}$ the apparently clear fluid rotating round the cells of vallisneria is resolved into multitudes of extremely minute particles of colourless matter, or bioplasm, every one of which possesses the power of moving, and is alive. There is, then, nothing improbable in the supposition that minute germs might pass through the cell wall with the pabulum. They would remain in the cell wall or tissue perfectly inactive and dormant as long as the cell remained vigorous and healthy, but sooner or later, if not from disease, from old age, changes must occur by which a state of things results which is favourable to the germs whose turn invariably comes at last. These grow and multiply and live upon the dead germinal matter and the altered and softened formed material of the cells. We must, in the absence of positive demonstration, hesitate to accept the doctrine that the lowest organisms may result from degradation of the living matter, which at one time formed a part of a higher being. But we may *refuse* to accept the statements which have been made as to the direct conversion of the fibrillæ or discs of striped muscle into bacteria, because such assertions are contradicted by well-known facts. From the red blood corpuscles may be made bodies which might be mistaken for bacteria; nay, if the most practised observer were to examine one of these bodies only by itself, he might easily be deceived. The little beaded filaments exhibit movements which, though differing from the movements of the bacteria, would certainly be mistaken for them by an unpractised observer. But it ought not to be necessary to state, that in microscopical research mere resemblance in external form and general characters should not be accepted as proof of identity of nature any more than in ordinary observation.

Of Diseases known to be due to Vegetable Organisms.—The diseases of man and the higher animals, known to depend upon the growth and development of vegetable organisms, are local affections confined to a part of the body not involving the blood, while for the most part, the different forms of contagious fevers are general affections in which the whole mass of the blood, and, in some cases, every part of the body, is affected, and is capable of communicating the disease. In fungus diseases, the structure of the vegetable organism can be made out without difficulty, and the vegetable examined in every stage of its development. The microscopic characters are distinct and definite enough. No such success attends our efforts to prove that vegetable organisms are truly the active agents in contagious fevers. And in many of the diseases which are at this time considered to be actually due to the multiplication of vegetable germs, it is doubtful if the tissues and organs invaded were perfectly healthy at the time of invasion. For all persons exposed are not attacked, and if not in all, at least in the great majority of instances known, the view that a morbid change must occur before the tissue is in a state to be invaded by the fungus growth, is tenable. In fact, it has been already shown that the fungi which commonly grow on the surface, and in other parts of the body, do not *produce disease*. The germs of fungi may remain perfectly passive and quiescent in healthy textures, growing and multiplying only in those which have already deteriorated in consequence of disease or old age. The growth of the vegetable germs, therefore, instead of occasioning the disease, may be dependent upon the occurrence of phenomena altogether different. There are, I think, very few morbid conditions that are unquestionably solely due to the growth and multiplication of vegetable fungi.

Some difficulties which prevent us from accepting the Vegetable Germ Theory of Disease.— If contagious diseases are due to the entrance into the organism of such minute vegetable germs as those described, is it not wonderful that any one escapes disease? Multitudes of germs of different species, as numerous as are the contagious diseases from which we suffer, must, if this theory be true, surround us. And yet the fungus germs, which are to be detected easily enough, and which indeed do exist in great numbers, are not known to cause any disease. Still, upon this view these must be the disease-producing particles, for they are the only vegetable germs that have been discovered. Passing into our lungs with every inspiration, entering our stomachs with our food and drink, everywhere in contact with our cuticle, in the chinks of which they might grow and multiply, these fungus germs must, one would think, pass in vast numbers into our blood, and be carried to every part of our bodies. Contagious diseases ought, therefore, to be more common than they are, and escape from attack should be almost impossible.

Vegetable fungus germs are to be met with in every

country, and there are probably few substances in or upon the earth which are entirely free from them. If their introduction alone is sufficient to produce disease, one malady ought to follow another, until the catalogue of contagious diseases becomes exhausted, or the organism is destroyed. But many fungi even form articles of diet and medicine, and many animals devour whole forests of living, growing fungi in every mouthful of food they take. Of these not a few are destroyed by the fluids poured into the alimentary canal, digested, and the products appropriated by the organism. The animal, in fact, lives upon them, instead of the fungi living upon him; and in various cases in which certain fungi do actually invade our tissues, the evidence of change in these last having occurred prior to the development of the fungi, is sometimes so distinct, that the conclusion is irresistible, that, so far from the fungus attacking a healthy structure and damaging it, the structure itself had deteriorated and changed, or had undergone morbid derangement ere it was invaded. By decay it would appear that it had become converted into material adapted for the nutrition of the fungi, the growth of which had been effectually resisted as long as the tissue remained healthy. If this be so, the fungi cannot be regarded as the *cause* of the disease, any more than the vultures which devour the carcase of a dead man can be looked upon as the cause of his death.

Vegetable germs exist in countless multitudes

where contagious diseases are unknown, as well as where they are rife. Their sparing or abundant multiplication varies with altering temperature, moisture, and other conditions, and does not always coincide with the fluctuation of disease. If vegetable germs are the seeds of disease, the seeds are everywhere, while in many instances the diseases are remarkable for being particularly local. If these be disease germs, they are present in all climes, while the diseases themselves are limited to certain definite regions. We may cultivate the vegetable germs without producing disease, and disease may be raging while there is no evidence of a corresponding increase of the vegetable organisms upon which it is supposed to depend. If vegetable organisms are really the contagious particles, it is hopeless to attempt to protect ourselves from their invasion, and to talk of extirpating them would be absurd, for were a particular species destroyed over half England to-morrow, the next breath of wind would bring multitudes of germs to take the place of those which had been swept away. Nor should we stand any chance of escaping their ravages, by leaving our dwellings in cities, and taking up our abode in the country, or by taking refuge even in the highest mountains, or other sequestered places far away from the haunts of men. And if fungi are developed spontaneously, and disease germs consist of fungi, the state of things is still worse, as in that case, if eradication were possible, it would be idle to attempt to effect it; for if all in existence at any one time were utterly destoyed, new ones would soon spontaneously emanate from the non-living, and we should be in as bad a plight as before. Minute vegetable germs, resembling those to which disease has been attributed, are everywhere, though their presence may easily escape observation. If, however, the pabulum adapted for them be present, and the conditions favourable to their development, they soon grow and multiply, and abundant evidence is afforded of their presence.

In answer to the observation, that if these fungus germs constitute the morbid material of contagious diseases everyone should be attacked, it might be said, "the organism is not always in a state favourable to invasion, and that it is only in exceptional cases, or in exceptional states of health, that the presence of fungi affects us deleteriously." To this the reply might be that "there are many kinds of contagious matter which give rise to characteristic effects with unerring certainty." The introduction of as much as would adhere to a needle point into the body of a healthy subject, acting without a chance of failure. If, therefore, we accept this vegetable germ theory of disease, we must hold that there are certain fungi which affect all men in all conditions of health, but which are at present undiscoverable, while other fungi, which are very easily discovered, are not known to affect the organism in any condition of health;

and that other fungi, also unknown at this time, exist, which are only able to produce their effects in organisms changed by certain previous actions for their reception,—and this, in spite of the fact that no connection whatever has been shown to obtain between any contagious disease and any kind of fungus germs.

But yet in favour of such a doctrine it might be urged with truth, that some *parasitic organisms* affect all indiscriminately, while others require certain preliminary changes to be carried out before the various parts of the organism they delight in are adapted for their habitation and are rendered favourable to their increase. It must not, however, be forgotten that parasites which are known exhibit at one or other stage of existence certain well-marked characters by which they may be recognised with the utmost certainty, and this is especially the case with parasitic vegetable organisms, many of which can be grown artificially without much difficulty, and studied in the several different stages of their development.

Those who look with partiality upon the vegetable germ theory of disease should consider how the absence of any bodies like vegetable fungi in animal fluids and solids, proved by experiment to possess active infectious properties, is to be accounted for. Not only is it the case that vegetable organisms are not to be found in the perfectly fresh virus when it is most active, but no specific form of vegetable growth can be developed from the particles which do exist, as would almost certainly be the case if the particles present in the fluids had been vegetable germs. Every kind of parisitic germ known is capable of undergoing development into a body having definite and well-marked characters. Though in the germ stage different species would resemble one another, as indeed is the case as regards creatures much higher in the scale, they do not constantly retain indefinite characters. And when the germs are so minute as to be readily passed over in ordinary microscopical examination (100 to 300), by the aid of higher powers excessively minute vegetable germs may be recognised with certainty, if not by their form, at least by their mode of multiplication. The germs of many animal parasites are also to be distinguished by careful examination, and from what we know of the life history of these, we should not be justified in attributing contagious diseases, in which every drop of animal fluid in the body possesses contagious properties, but for a fixed and definite period of time only, to germs of a new class of animal or or vegetable parasite of which not one species has been discovered, and the germs of which are even less than $\frac{1}{100000}$ of an inch in diameter. It may, therefore, be affirmed that the matter which forms the active virus or poisonous material does not exhibit the properties of any vegetable or animal parasitic organism yet discovered and identified. Neither can any organisms, having special and peculiar characters, be developed from any definite virus.

Will, then, the advocates of the vegetable germ theory of disease maintain that this view ought to be accepted simply because, in some of the discharges and fluids of diseased animals or man, vegetable germs are to be found, in face of the fact that similar germs are to be detected in all sorts of *harmless animal fluids and even in foods which are taken and digested!* As soon as fungi have developed themselves freely in animal fluids possessing special contagious properties, such as vaccine lymph, or smallpox lymph, the specific characters of the poison become weak or disappear. This seems to negative the view under consideration. In answer it might be urged that, "because a few vegetable organisms excite the disease, it does not therefore follow that a multitude should be more potent,—rather the contrary; for a few might retain their vitality and propagate themselves, while, if a great number were present, the pabulum necessary for their activity would be insufficient, and all would perish!" The advocates of the theory may be permitted to enjoy any advantage that can be derived from this sort of argument; for, however cleverly it may be put forward, most people who know the facts of the case will be of opinion that the vegetable organisms when present are but accidental concomitants, and that a potent poison, not of the nature of a vegetable germ, is present in the animal fluid or solid in which the contagious properties are known to reside.

www.ingramcontent.com/pod-product-compliance
Lightning Source LLC
Chambersburg PA
CBHW020901160426
43192CB00007B/1027